CAMBRIDGE COUNTY GEOGRAPHIES

SCOTLAND

General Editor: W. Murison, M.A.

LANARKSHIRE

Cambridge County Geographies

LANARKSHIRE

by

FREDERICK MORT, M.A., B.Sc., F.G.S.

Fellow of the Royal Scottish Geographical Society
Late Lecturer in Geology, Glasgow University

With Maps, Diagrams and Illustrations

Cambridge:
at the University Press
1910

CAMBRIDGE UNIVERSITY PRESS
Cambridge, New York, Melbourne, Madrid, Cape Town,
Singapore, São Paulo, Delhi, Mexico City

Cambridge University Press
The Edinburgh Building, Cambridge CB2 8RU, UK

Published in the United States of America by
Cambridge University Press, New York

www.cambridge.org
Information on this title: www.cambridge.org/9781107616707

First published 1910
First paperback edition 2012

A catalogue record for this publication is available from the British Library

ISBN 978-1-107-61670-7 Paperback

CONTENTS

ILLUSTRATIONS

MAPS

Note. For the photographs on pp. 18 and 19 the writer is indebted to J. W. Reoch, Esq.; for that on p. 74 to Messrs Beardmore, Parkhead Forge; for that on p. 76 to the Manager, Hyde Park Works; for permission to reproduce the photograph on p. 34 to Prof. Watts, Secretary of the Geological Photographs Committee, British Association; and for permission to have the Lesmahagow Flagon photographed, to the Glasgow University Court.

1. County and Shire. The Origin of Lanarkshire.

The present sub-divisions of Scotland are the result of a long process of adjustment between different competing systems. The King, the Church, and the Nobles were centres of segregation that tended to group the community in different ways. Yet among these discordant forces the powerful influence of the natural, physical features of the country can often be seen to have shaped the political divisions in harmony with natural regions. Of this fact there is no better example in Scotland than the county of Lanark. The modern county is the division of the kingdom administered by a sheriff, and this system dates back at least as far as the reign of David I (1124–1153). When the crowns were united in 1603, the districts administered by the sheriffs of the king coincided with the modern counties, except that Caithness, Sutherland and Ross were under the jurisdiction of the Sheriff of Inverness.

The shires, that is *shares*, were originally governed by the great earls of the country, who in many cases took their titles from the districts they ruled. When William I had conquered England, many of the English earls were dispossessed of their lands, which were given to William's companions or *comites*. Each district was therefore called a *comitatus*, from which we get the word *county*. The English feudal system was introduced into Scotland by David I, and the sheriffdom or county of Lanark probably dates from his reign. William Hamilton, of Wishaw, writing about 1710, tells us that "The shyre of Lanark was anciently of greater extent than now it is; for there was comprehended in it the whole sheriffdome of Ranfrew, lying laigher upon Clyde...untill it was disjoyned therefra by King Robert the Third, in anno 1402." Since that time the changes in the county boundary have been geographically unimportant. Many meanings of the word Lanark have been suggested, but most authorities are agreed that it is derived from *llanerch*—a clearing in a forest, a word belonging to the Welsh or Cymric branch of the Celtic group of languages.

It has already been stated that Lanarkshire is a good example of the way in which natural physical features have influenced the political divisions of a country. For the county is a geographical unit, namely, the basin of the Clyde, a fact that is well expressed by the old name Clydesdale. The most southerly part of the shire is Gana Hill, and on the slopes of this hill the Clyde rises. To the north-west the county ends just where the river becomes too wide to be bridged or crossed conveniently. Thus while Lanarkshire embraces both banks of the Clyde, nearer the sea the broader river forms the boundary between Dumbartonshire and Renfrewshire. Naturally the limits of the county do not everywhere coincide exactly with the watershed of the Clyde, yet for considerable distances the county boundary is absolutely identical with the watershed of the river.

2. General Characteristics. Position and Relations.

Imagine two lines drawn from north-east to south-west across Scotland, one from Stonehaven to Helensburgh, the other from St Abb's Head to Girvan. These two lines divide Scotland into three districts called respectively the Highlands, the Central or Lowland Plain, and the Southern Uplands; and these three districts differ strongly in physical aspect, in rocks, in scenery, in vegetation and in industries. The lines mark the course of two great faults or cracks, which traverse the whole country, and between which the land has gradually sunk for thousands of feet. This sinking of the central part of Scotland took place many ages before man inhabited this country, but yet it may be considered the most significant stage in the evolution of Scotland, for it preserved the all-important coal fields of the lowlands on which the prosperity of the country largely depends.

The Central Plain of Scotland is not only the most fertile part of the country, but by far the greatest proportion of the mining and the manufactures is carried on there. It has thus become a district unique in Great Britain, for it corresponds to no less than four separate parts of England—the south-eastern plain devoted to agriculture, the Black Country with its coal and iron industries, the woollen district of Yorkshire, and Lancashire with its cotton manufactures. In some respects Lanarkshire is the most typical county of the Lowlands, and its diversity of surface and variety of industries are increased by the fact that the most southerly of the two great faults mentioned crosses the Clyde near Roberton, so that Lanarkshire is partly in the Lowland Plain and partly in the Southern Uplands.

The position of Lanarkshire on the western slope of the country was at first a disadvantage. For long the eastern coastal strip was by far the most important part of Scotland. The commerce of Europe to a large extent was carried on in the districts bordering the North Sea. In Lanarkshire, however, just as the drainage of the county is collected by a thousand little streams that feed the main current of the Clyde, sweeping down with ever increasing volume to the western sea, so the movement of trade naturally tended to take the same course. The face of the county was turned away from the chief commercial centres, but the progress of civilisation in its westward march has shifted the balance of trade to the shores of the Atlantic, and thus the geographical position of Lanarkshire at the present time is one of its most important advantages. The stream of trade that now pours out of Lanarkshire to the Atlantic is double that from all the other counties of Scotland put together.

Position alone, however, does not fully explain the phenomenal growth of the county during the last century and a half. Its natural resources are just as important. Its mineral wealth surpasses that of all the other counties of Scotland, and it may be said with truth that Lanarkshire owes its importance to coal and iron. In the latter half of the eighteenth century, steam power began to be applied to the world's industries. Coal was urgently needed, and for the first time coal could be easily obtained, for it was not till James Watt had improved the steam engine that adequate means were available for pumping water from the mines. The introduction of machinery necessitated a great increase in the production of iron, and this reacted again on the coal trade in the demand for coal for smelting purposes. Scores of busy industrial centres sprang into existence in Lanarkshire, and soon the population far outnumbered that of any other county. The banks of the Clyde with easy access to the sea and their proximity to the coal and iron fields formed an ideal home for the shipbuilding industry. The yards of Renfrewshire and Dumbartonshire launch many a goodly ship, but Lanarkshire supplies the material.

3. Size. Shape. Boundaries.

Although Lanarkshire contains more than a quarter of the population of all Scotland, in size it ranks only tenth among the counties. (See Figs. 1 and 2, p. 167.) Inverness is nearly five times larger than Lanark, but the latter has nearly fifteen times the population of the former. From north-west to south-east the shire extends for fifty miles, while a line drawn east from the point where Dumfries, Ayr and Lanark meet to the eastern edge of the county stretches for thirty miles. The total area, including water, is 567,385 acres. The county is fairly symmetrical in shape. As the basin of the Clyde broadens on going down the river from its source, so the county widens to a maximum and then narrows towards its north-western extremity. Slight changes were made in the boundary in 1891 and 1892 by the Boundary Commissioners, chiefly to correct the anomaly of a parish being partly in one county and partly in another.

Starting from Gana Hill in the south, the boundary strikes north-north-west, coinciding exactly with the watershed between the Nith and the Clyde, and passing over hills about 2000 feet in height. A little north of Slough Hill the boundary swings to the west and leaves the watershed, but only for about two miles. Keeping to the watershed again, the boundary line runs west over hills about 1500 feet in height till it reaches Threeshire Stone, the point where Ayr, Dumfries and Lanark meet. The line now runs along the watershed between the Clyde and the Ayr, passing Cairn Table on the way, and then swings away west in order to include the head-waters of the Avon. Leaving the watershed the boundary now follows the Avon for some distance and then goes north-north-west to the point where Ayr, Renfrew and Lanark meet. Bending north the line follows the Cart for some distance, then leaving the Cart it reaches the Glasgow boundary, follows it by bending to the west, and so runs down to Renfrew through Govan. From Renfrew the line turns east and crosses the Clyde just west of Whiteinch. Passing to the Kelvin the line follows that river almost to Kirkintilloch, where it turns south and then east through Lenzie to the Luggie, a tributary of the Kelvin. Leaving the Luggie the boundary crosses the watershed of the country to the Avon, a tributary of the Forth. Then it turns to the south-east by Black Loch and next bends sharply east till it is only three miles from Bathgate. It runs south now to Black Hill, then bending north-east it rejoins the watershed at Leven Seat. It cuts across the south part of Cobbinshaw Reservoir and then follows the course of the Medwin, keeping just to the east of the watershed. It reaches the watershed again at Broomy Law, but then keeping to the east of Biggar the line encroaches on the Tweed basin, a strip of which is included in Lanarkshire. It comes back to the watershed at Scawdmans Hill, and follows it south to Clyde Law by hills over 2000 feet in height. Passing the point where Lanark, Dumfries and Peebles meet, the boundary coincides with the watershed between the Annan and the Clyde, and runs in a zig-zag line through Lamb Hill to Earncraig Hill, finally swinging sharply round to Gana Hill again. The detailed tracing of the boundary has shown us, therefore, that on the whole Lanarkshire may be fairly described as the upper and middle basin of the Clyde, and the modern county is almost equivalent to the mediaeval division of Clydesdale.

4. Surface and General Features.

The surface of Lanarkshire is extremely varied, ranging from a height of nearly 2500 feet down almost to sea level. As has been mentioned, the south-east portion of the county forms part of the Southern Uplands, and the highest hills are almost exclusively confined to this district.

Leadhills. Behind are the "Lowthers," the highest hills in Lanarkshire

With the exception of Tinto (2335 feet), all the hills above 2000 feet are found south-east of a line drawn through Culter, Lamington, Roberton and Crawfordjohn. A glance at the physical map on the cover will show at once the strong contrast between the districts on each side of the line mentioned.

The highest hill situated entirely in the county is Green Lowther (2403 feet), two miles south-south-east of Leadhills, although Culter Fell on the Lanark-Peebles border is 50 feet higher. Green Lowther forms one of a group of hills over 2000 feet in height, which run in a north-east and south-west direction from the Clyde to the county boundary, and of which the next in altitude is Lowther Hill (2377 feet). From the names of its two highest summits the group is sometimes called the Lowther Hills. There are in all about a score of hills in Lanark two thousand feet or more in height, of which about one half are entirely in the county, the others being on the boundary line. Of these, in addition to the hills mentioned, the most important are Glenwhappen Rig (2262 feet) on the Peebles border, Rodger Law (2257 feet) and Ballancleuch Law (2267 feet) in the southern extremity of the shire, and Dun Law (2216 feet), one of the summits of the Lowther group. The Tinto Hills are the most conspicuous in the county, as they are separated from any other important group. Tinto Tap is a landmark all down Clydesdale, and it is said that from its summit on a clear day parts of no fewer than sixteen counties can be seen.

The upper part of Clydesdale has a charm of scenery that is confined to the Southern Uplands. The district does not make the same sudden and arresting appeal to the unobservant traveller that some parts of Scotland do. The wildness, ruggedness and grandeur of the Highlands as a rule are absent, for the outlines of the hills are generally smooth and rounded, yet there is a softly flowing sweep of contour, a tenderness of colour and a melancholy loneliness about these green and treeless summits that make a quiet but irresistible appeal to the wanderer among them.

In Lanarkshire the transition from the Southern Uplands to the Lowland Plain is not so abrupt as it is farther to the north-east or farther to the south-west.

Characteristic scenery of Southern Uplands. View across the Clyde valley near Crawford

The higher parts of the Lowlands, particularly away from the Clyde valley, are bare and bleak moors undulating monotonously almost as far as the eye can reach. The land rises not only to the south, but also as it recedes from the Clyde, so that the highest parts form the boundary of the county. Thus the more fertile central part is flanked by long stretches of barren moorland, useless for agriculture, but in many cases forming good shooting districts. Yet these moorlands are intersected at intervals by unsuspected glens of rare beauty. Although the rocks and therefore the details of the scenery are different, the district in this respect has a strong resemblance to parts of Derbyshire, where the same alternation of featureless uplands with sudden bits of charming river scenery is found. Many of the smaller streams of the county, almost unknown except to those living in the locality, will bear comparison with the finest parts of the Clyde itself.

From the Falls of Clyde to Bothwell the scenery is almost uniformly beautiful. The bareness of the Southern Uplands is gone. The river flows through a green and fertile country, well-wooded and dotted over with fine mansions. The valley is broad with gently shelving banks, although in places it contracts and takes on the character of a gorge. The lower ward of Lanarkshire is somewhat flat and unpicturesque. The most noticeable height is the ridge that runs parallel to the Clyde from Cathkin to Dechmont.

5. Watershed. Rivers. Lakes.

We have already shown that the watershed of Lanarkshire coincides approximately with the county boundary. The watershed of the southern half is a framework of high hills shaped like a great irregular V, with the point at Queensberry Hill just south of the county boundary. On the north-east side while the boundary follows the general direction of the watershed, it swings first to one side and then to the other. There is here no well-marked line of hills, but a wide expanse of bare and lonely moorland, eight or nine hundred feet above the level of the sea.

There is a prevalent but mistaken belief that a watershed must be a range of hills, or at any rate must stand well above the level of the surrounding country. In many cases this is not so. The watershed may be a flat marsh, and one may sometimes walk right across an important watershed without noticing any change of slope whatsoever. This is illustrated in an interesting way by the Clyde near Biggar. At this point the main river actually is within a mile of the watershed between the Clyde and the Tweed. The divide is the broad flat valley of Biggar Water, and in times of heavy flood the waters of Clyde and Biggar mingle. It would be an easy task to divert all the head waters of the Clyde at this point into the North Sea. In fact, Michael Scott, the famous warlock, is said to have been in the act of doing this. He was marching down the vale of Biggar with the Clyde roaring at his heels when he became alarmed at the threatening sound behind him. Fortunately for the present prosperity of Glasgow he looked back, the spell was broken, and the waters resumed their usual course.

The explanation of the origin of this low pass between the basins of the Clyde and the Tweed is not easy. Sir Archibald Geikie attributes it to "the recession of two valleys and to the subsequent widening of the breach by

Falls of Clyde—Bonnington Linn

atmospheric waste and the sea," but this is not convincing. An explanation has been given that is more probable, although it involves a startling readjustment of our ideas regarding the permanence of natural features such as rivers. On this theory the Clyde now flows in an opposite direction to its course in former times. Originally it took its rise somewhere on the western border of Scotland, at least as far west as Loch Fyne, at a time when Loch Long and Loch Lomond were not in existence. Like the Tay and the Forth and the Tweed, it flowed south-east to the North Sea, cutting the valley where the Biggar Water now runs. Later on when Loch Long and Loch Lomond were formed, they cut across the original head-streams of the Clyde, diverting them to the Atlantic and leaving a dry valley where the modern "Tail of the Bank" is situated. This was occupied by a westward flowing stream that rapidly thrust its head backwards and occupied the old valley, thus becoming the parent of the present Clyde. This explanation, fanciful though it seems at first sight, has many facts in its favour; and other passes similar to that of Biggar, but in different parts of Scotland, have been explained on the same principles.

Though not the largest river in Scotland, the Clyde is by far the most important. It is the gateway to one of the great industrial districts of the world. Along its banks more than one-third of the total population of Scotland is clustered. No other river in Britain can show such strange and violent contrasts in its course as are revealed by a walk along the valley of the Clyde from source to sea. From its source for many miles its course is through the Southern Uplands, a pure mountain stream among lonely hills, still clear and unpolluted as when the Roman legions tramped along the old road by its banks. After issuing from the hills, it flows through bare moors till it enters the ravine above the falls. It races through its gorges and leaps its falls as if in haste to reach the garden of the Clyde, the orchard belt,

> "The pleasant banks of Clyde
> Where orchards, castles, towns and woods
> Are planted by his side."

The sternness and bleakness of its upper course have vanished. The landscape seems hardly Scottish in its rich, luxuriant beauty. But the pall of smoke is already visible in the west, and soon the Clyde flows through the "black country" of Lanarkshire, almost every town along its banks eager to defile its purity with every conceivable form of industrial waste and pestilential sewage. From here to its mouth the Clyde is a slave to commerce; and foundries, mills, engineering-shops and shipyards roar about its banks till it escapes at last and finds rest in the clean, salt waters of the firth. Beautiful though the river is in its upper reaches, surely there are no lovelier scenes in Scotland than on the estuary of Clyde.

According to tradition the source of the Clyde is on Clyde Law, down the slopes of which runs Little Clydes Burn. The older writers are unanimous in this opinion:

> "Annan, Tweed and Clyde
> Rise a' out o' ae hill side."

But if we seek, as we should, for the true head-waters in the most important stream, we must select Daer Water, rising at a height of 1600 feet above sea-level on the slopes of Gana Hill on the southern border of the county. Below the point where the Daer Water joins the Powtrail Water, the united stream is called the Clyde by the Ordnance Survey, the supreme authority on matters topographical. The valley of Daer Water is perhaps the most inaccessible part of Lanarkshire. There is no road within many miles; a little footpath only runs up the valley, deserted save for a shepherd or at infrequent intervals a solitary fisherman.

A mile and a half past Watermeetings is the junction with Clydes Burn, and down its valley from Beattock Summit come the road and the railway from the south to keep the Clyde close company all the way to Glasgow. The engineers of road and railway knew well the easiest and straightest way from Carlisle to the rich Scottish Lowlands; but long before their time the Romans, with unerring skill, had discovered the route up Annandale, across Beattock Summit and down Clydes Burn to Clydesdale.

Still north the river flows receiving Elvan Water, where the gold-seekers even yet search for specks of the precious metal in the sands and gravels. At Crawford the river swings west for a mile before resuming its northward course. On the right bank is Tower Lindsay, the ruined stronghold of the Lindsays, Earls of Crawford. With its inseparable comrades, the railway and the road, on each side, the river flows to Abington, where in 1839

Prince Louis Napoleon, who was to sit on the throne of France, took his supper by the kitchen fire of the little inn. Duneaton Water comes in on the left bank from far Cairn Table and the Ayrshire border, and then the stream flows past Roberton and leaves the Southern Uplands behind it.

The Clyde now takes a bend to the east that brings it to the head of Biggar valley. Tinto—the hill of fire—dominates all this part of Clydesdale. The commanding appearance of Tinto from almost any part of the Clyde valley is due to its splendid isolation, far from any rival peak. It heaves a huge shoulder, curving to a massive dome, for nearly 2000 feet above the level of the surrounding district, and from its summit can be seen a wide expanse of country, from the Bass Rock to Arran, from the Grampians to the peaks of Cumberland. On its south-east slope are the remains of Fatlips Castle, overlooking Symington.

Northwards again the stream flows, now a stately river 40 yards in width, among its fertile haughs, past Thankerton and Covington. On receiving the Medwin the Clyde turns first to the west and then to the south-west, making the curious curve in its course that finishes just before the falls. Here comes in the Douglas Water from Douglasdale, home of the most powerful family that ever lived in Scotland, greater often than the kings themselves. Where the Douglas joins the Clyde the river is flowing gently between sloping banks, while all around for many miles the ground is covered with great heaps and ridges of sand and gravel, the remnants of the Great

Ice Age. Soon the valley contracts, the speed of the river increases, it hurries breathlessly down over a series of rapids and then with a roar makes its first leap over Bonnington Linn. Below Bonnington the scene is magnificent. The river toils and foams along in a deep gorge walled in by rocky cliffs 60 feet in height, in many places beautifully clothed with foliage. For half a mile

Gorge above Corra Linn

the chasm continues till at a sharp bend in the river, the water leaps in a mass of foam with a noise like thunder over Corra Linn. The river valley below the fall is a veritable cañon; the sides are dark precipices over 100 feet in height, amazing not only to the actual eye but to the mental vision that sees that this defile has been caused by the gradual recession of the falls up stream.

Below Lanark the Mouse Water comes in through the high and narrow defile of the Cartland Crags, "dark, rugged and precipitous crags, which are the astonishment

Falls of Clyde—Stonebyres Linn

and terror of every beholder," according to the writer of the *Statistical Account*. Our nerves are surely stronger now than they were a hundred years ago. Still the gorge of the Mouse here is certainly magnificent, and one can hardly realise that one is standing by the same stream that a few miles higher up steals along so gently through the flat, bog country round Carstairs. Stonebyres Linn, a mile below the junction of the Mouse Water, is the last of the falls, in time of flood an unbroken drop of 70 feet. Of the three falls (for Dundaff Linn is hardly worth considering compared with its greater sisters) Corra Linn has the greatest reputation, and is by far the most visited. Yet the others are hardly, if at all, inferior. Stonebyres in particular, from certain points of view, is perfect, and worth more attention than it receives.

The origin of the falls is an interesting problem. It is supposed that the course of the Clyde near Lanark before the Great Ice Age was quite different from what it is now. The curious curve from Hyndford Bridge to Lanark did not exist, the river taking a fairly straight course between these points. The Douglas Water also had a different course, as is shown in the map on p. 21. The old channels were filled up with material brought by the glaciers, and after the melting of the ice the former channels were not everywhere re-excavated. When the grip of the ice was released, the Clyde joined the former channel of the Douglas near the site of Core House, and easily scooped it out, forming a waterfall over the old bank. The waterfall thus caused receded, as all waterfalls do, until it reached its present position at Bonnington Linn. Where the Clyde joined its old course near the site of New Lanark another fall was formed that worked backwards, and is now known as Corra Linn. Stonebyres

Linn seems to have been caused in quite a different way. A very hard conglomerate full of quartzite pebbles crosses the river about 300 yards below the present position of the fall. The softer beds under this were eroded more rapidly than the hard bed, and so a waterfall was formed. It is in this way that most waterfalls have been formed.

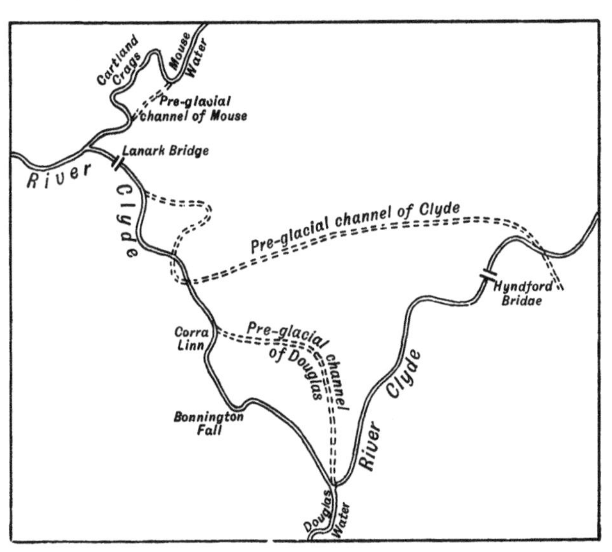

Sketch Map illustrating the origin of the Falls of Clyde
(After Stark, *Trans. Geol. Soc. Glasgow*)

Half a mile from the right bank of the Clyde stands Lanark. Near here, if tradition is correct, took place the first serious encounter of Sir William Wallace with the English. Incensed at a jest against his young wife, Wallace drew his sword and cut off the hand of the offending Englishman. A general fight took place, but the English garrison poured out in overwhelming numbers, and Wallace escaped from the town through his own door opened by his wife. For this his wife was slain, and Wallace vowed a lasting enmity to the English. With a few brave followers he attacked the garrison and slew many of them, including Hazelrig the governor.

From Lanark down to Bothwell is the orchard country famous throughout the centuries since the time of Bede. In May the valley is white with blossom. Fruit trees and currant bushes clothe the slopes to the very edge of the river, and send fragrant offshoots up every tributary stream. Strawberry culture, though a recent introduction, is outstripping the other fruits; and still more recently ugly little tomato houses seem to have sprung up everywhere from the ground to meet the rapidly increasing demand. On the right bank we look up at the entrance to the fine gorge of Fiddler's Gill, and almost opposite the Nethan comes in, with Craignethan Castle on its left bank. Several miles lower down, the Clyde is joined by the Avon Water, a fine stream for trout and grayling. Not only in fish does it rival the Clyde, for its scenery in some parts is equal to anything the main river can show. Modern mansions and castles old in story are round us, but for the present must be left behind. There is a darkening haze to the west, and at times we catch a glimpse of tall chimneys warning us of a different type of scenery if we leave the river's banks. At night the indications of Lanarkshire's "black country" are even more apparent. Alexander Smith has well described the impression a traveller to Glasgow obtains who approaches the city from the south at night.

> "The wild train plunges in the hills,
> He shrieks across the midnight rills;
> Streams through the shifting glare
> The roar and flap of foundry fires,
> That shake with light the sleeping shires;
> And on the moorlands bare,
> He sees afar a crown of light
> Hang o'er thee in the hollow night."

The proximity of the great industrial centres gives a peculiar atmospheric effect that is finely portrayed in a little word-picture from the pen of Sir Harry H. Johnstone in his *Life of Livingstone*, a description remarkable both for vividness and accuracy. "Beyond the factories, with the invisible Clyde rushing over weirs in the gorge between, is a high ridge of wooded down; and above all that strange, opalescent heaven, with its rainbows and curtains of vapour, its wreaths and rolling masses of cloud, its mists and films of smoke, its watery sunshine or its livid glare of fire, when night falls and the smoke-pall which daylight has rendered so dull-coloured and opaque becomes one vast shimmer of rosy flame."

The lower part of the course of the Clyde has always been subject to destructive floods. In 1454 "ther wes ane right gret speit in Clyde, the quilke brocht down haile housis, bernis and millis," and even yet at intervals much damage is done to crops and houses by the river overflowing its banks. An inundation in 1831 caused irreparable loss by the destruction of a large number of letters written by Robert Burns to his friend William Reed, the publisher.

The valley here is open and the river winds through rich haugh lands, but below Bothwell Brig the banks close in and steepen with a marked change in the character of the scenery, which is here exceptionally fine. Two miles below the bridge old Bothwell Castle looks across to Blantyre Priory on the other bank, comrades for nigh six hundred years. A little further down, at Kenmuir and Carmyle, the Clyde is linked with the names of many famous artists. Sam Bough, McWhirter, Horatio McCulloch, and many others of lesser note have painted on these banks. But a forest of chimneys, a wilderness of stone and lime is near at hand, and after doubling from one side to the other time and again as if looking for some way of escape, the river glides slowly into the heart of Glasgow.

From Glasgow to the firth the Clyde is largely the product of man. The conversion of a stream, in places but a few inches in depth, into a water-way for ocean-going ships is one of the romances of industrial history. In the sixteenth century an attempt was made to improve the channel at Dumbuck but was not successful. The magistrates of Glasgow therefore reported in 1668 that they had had "ane meeting yeasternight with the lairds, elder and younger, of Newark, and that they had spoke with them anent the taking of ane piece of land of theirs in feu, for loadning and livering of their ships there, anchoring and building ane harbour there, and that the said lairds had subscryvit a contract of feu this morning:

quhilk was all allowed and approvine be said magestratis and counsell."

On the ground thus purchased the magistrates laid out the town of Port Glasgow with harbours and a graving dock. Here the goods were taken from the ships and loaded on the backs of little pack-horses that brought them by badly made tracks to Glasgow. In 1755 the river was still in a state of nature, for between Glasgow and Renfrew there were twelve shoals, one of which was only 15 inches deep at low water. James Watt surveyed the river in 1769, and reported a depth of 14 inches at Hirst Ford during low water. To John Golborne of Chester is due the first marked improvement in the navigation of the river, which was dredged and also narrowed by the construction of jetties. A few years later Golborne deepened Dumbuck Ford to a depth of seven feet, and owing to the scour of the river due to his jetty system this depth was in 1781 found to have become 14 feet. Act after act was carried through Parliament giving new powers, and each meant a further improvement in navigation and a consequent stimulus to the commerce of Glasgow. A great advance was made by the application of steam power to dredgers and the adoption of steam hopper barges, to which the present state of the river is largely due. A formidable obstacle was found in the Elderslie Rock, extending right across the river at a depth of eight feet below low water. After years of labour this was removed at a total cost of about £140,000, giving now a depth at low water of 28 feet.

The principal tributaries can be easily remembered

from Wilson's enumeration of them in his poem *The Clyde.*

"Glengonar's dangerous stream was stained with lead;
Fillets of wool bound dark Duneaton's head;
With corn-ears crowned, the sister Medwins rose,
And Mouse, whose mining stream in coverts flows;
Black Douglas, drunk by heroes far renowned,
And turbid Nethan's front with alders bound;
Calder, with oak around his temples twined,
And Kelvin, Glasgow's boundary flood designed."

The lakes of Lanarkshire are neither numerous nor large. Hillend Reservoir, five miles north-east of Airdrie, is the largest, being about a mile long. It supplies the Monkland Canal with water. There is a little group of lochs north-west of Coatbridge, of which the largest is Bishop Loch. The others include Lochend, Woodend and Johnston Lochs. East of Glasgow is Hogganfield Loch, and near Lanark with wooded banks is Lang Loch. As a rule the loch fishing is poor, pike providing almost the only sport.

6. Geology and Soil.

The rocks are the earliest history books that we have. To those who understand them they tell a fascinating story of the climate, the natural surroundings and the life of a time many millions of years before the foot of man ever trod this globe. They tell of a long succession of strange forms of life, appearing, dominating the world, then vanishing for ever. Yet not without result, for each successive race was higher in the scale of life than those that went before, till man appeared and struggled into the mastery of the world.

The most important group of rocks is that known as *sedimentary*, for they were laid down as sediments under water. On the shores of the sea at the present time we find accumulations of gravel, sand and mud. In the course of time, by pressure and other causes, these deposits will be consolidated into hard rocks, known as conglomerates, sandstones and shales. Far out from shore there is going on a continual rain of the tiny calcareous skeletons of minute sea-animals, which accumulate in a thick ooze on the sea-floor. In time this ooze will harden into a limestone. Thus by watching the processes at work in the world to-day, we conclude that the hard rocks that now form the solid land were once soft, unconsolidated deposits on the sea-floor. The sedimentary rocks can generally be recognised easily by their bedded appearance. They are arranged in layers or bands, sometimes in their original horizontal position, but more often tilted to a greater or less extent by subsequent movement in the crust of the earth.

We cannot tell definitely how long it is since any special series of rocks was deposited. But we can say with certainty that one series is older or younger than another. If any group of rocks lies on top of another, then it must have been deposited later, that is it is younger. Occasionally indeed the rocks have been tilted on end or bent to such an extent that this test fails, and then we must have recourse to another and even more important way of finding the relative age of a formation. The remains of animals and plants, known as fossils, are found entombed among the rocks, giving us, as it were, samples of the living organisms that flourished when the rocks were being deposited. Now it has been found that throughout the world the succession of life has been roughly the same, and students of fossils (palaeontologists) can tell, by the nature of the fossils obtained, what is the relative age of the rocks containing them. This is of very great practical importance, for a single fossil in an unknown country may determine, for example, that coal is likely to be found, or perhaps that it is utterly useless to dig for coal.

There is another important class of rocks known as *igneous* rocks. At the present time we hear reports at intervals of volcanoes becoming active and pouring forth floods of lava. When the lava has solidified it becomes an igneous rock, and many of the igneous rocks of this country have undoubtedly been poured out from volcanoes that were active many ages ago. In addition there are igneous rocks—like granite—that never flowed over the surface of the earth as molten streams, but solidified deep down in subterranean recesses, and only became visible when in the lapse of time the rocks above them were worn away. Igneous rocks can generally be recognised by the absence of stratification or bedding.

Sometimes the original nature of the rocks may be altered entirely by subsequent forces acting upon them. Great heat may develop new minerals and change the appearance of the rocks, or mud-stones may be compressed into hard slates, or the rocks may be folded and twisted in the most marvellous manner, and thrust sometimes for miles over another series. Rocks that have been profoundly altered in this way are called *metamorphic* rocks, and such rocks bulk largely in the Scottish Highlands.

The whole succession of the sedimentary rocks is divided into various classes and sub-classes. Resting on the very oldest rocks there is a great group called Primary or Palaeozoic. Next comes the group called Secondary or Mesozoic, then the Tertiary or Cainozoic, and finally a comparatively insignificant group of recent or Post-Tertiary deposits. The Palaeozoic rocks are divided again into systems, and since the rocks of Lanarkshire fall entirely under this head, we give below the names of the different systems, the youngest on top.

Palaeozoic Rocks.

Permian System.
Carboniferous System.
Old Red Sandstone System.
Silurian System.
Ordovician System.
Cambrian System.

All these systems are represented in Lanarkshire except the oldest (Cambrian) and the youngest[1] (Permian).

A line running in a north-east and south-west direction by Crawfordjohn, Roberton and Lamington to the county border near Culter marks the position of a great crack or

[1] A small area in the Snar Valley between Crawfordjohn and Leadhills has been referred to the Permian system by some writers.

fault, to the north-west of which the rocks have subsided until the Old Red Sandstone System has been brought level with the Ordovician rocks that occur south-east of this line. The sudden change in the character of the country observed in crossing this line and already mentioned is thus due to the geological structure of the district. The Ordovician rocks consist of grits, conglomerates, flagstones, shales and cherts, along with volcanic lavas and volcanic ash. Since their deposition they have been thrown into such numerous and complicated folds that it is hopeless to determine from their present position what is the true order in which they were laid down. In a series of brilliant monographs, however, Professor Charles Lapworth, by a careful study of their fossils, demonstrated their true order of succession.

Along the crests of folds or arches in the Old Red Sandstone rocks north of the fault there are found bands of Silurian rocks. One band occurs near Douglas Water and another at Logan Water near Lesmahagow. The latter is noted for the peculiar forms of life that have been found there. Giant crustaceans and the very earliest known of Scottish fishes, some of them found nowhere else, have been collected there in great numbers by enthusiastic geologists.

The Old Red Sandstone rocks get their name from the fact that a considerable proportion, although by no means all of them, are sandstones of a red or brown colour. In Lanarkshire they occur in a very irregular band north-west of the fault separating the Lowlands from the Southern Uplands. This band extends down the Clyde to Crossford, and all the falls occur in the lower Old Red Sandstone. The rocks consist of conglomerates, grits, sandstones and mudstones, as well as rocks of volcanic origin. The igneous rocks stretch across the county from Douglas past Biggar to Dolphinton. At Tinto a great sheet of igneous rock has been thrust into the surrounding strata, and the origin of the Tinto Hills can be traced to the way in which this rock has resisted the ceaseless attacks of the weather. The roads in this district owe their red colour to the fact that the bright-pink igneous rocks of Tinto are the chief source of road-metal in the locality.

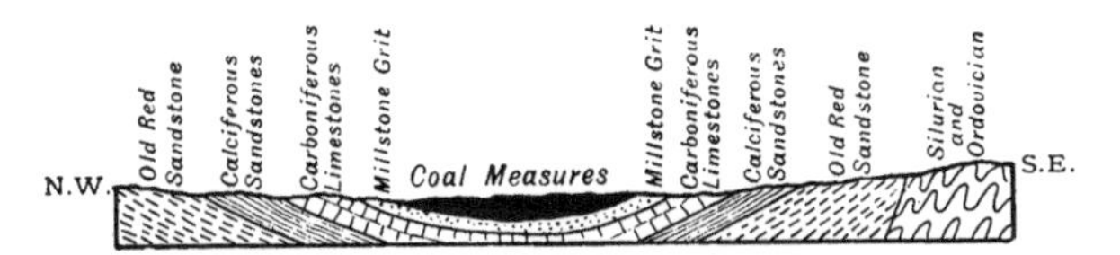

Section across the Lanarkshire Coal Basin

The Carboniferous System is by far the most important in the county, for it contains the Coal Measures on which the very existence of Lanarkshire as an industrial community depends. The sub-divisions of the Carboniferous System are as follows:—

Coal Measures.
Millstone Grit.
Carboniferous Limestone Series.
Calciferous Sandstone Series.

In Clydesdale these rocks have a synclinal or trough-like arrangement, so that the highest beds, the Coal Measures, appear in the middle bordered by lower and lower beds as we recede from them either to the north-west or the south-east. (See the section on p. 31.)

The Calciferous Sandstone series comprises conglomerates, marls, sandstones, shales, variegated clays and impure limestones. When these rocks were being deposited, volcanoes were active in many parts of Scotland. The high moors and hills that stretch for thirty miles north-west from Strathaven are formed of the lava that poured from the throats of countless volcanoes in early Carboniferous times.

The Carboniferous Limestone series consists of sandstones, shales, limestones, coal-seams and ironstones, and forms a belt surrounding the Lanarkshire coal-field on the north, west and south sides. These rocks are not the true Coal Measures, but many of the coal-seams have been worked. The valuable gas coal of Lesmahagow belongs to this division. Some of the other beds are also of considerable economic importance. The clayband and blackband ironstones, though now almost exhausted, have been extensively worked, as have also the numerous bands of limestone. In addition the sandstones furnish building material of the finest quality. A large proportion of the city of Glasgow has been built from the sandstones of this series that are found within a few miles of the centre of the town.

Overlying the Carboniferous Limestone series comes the Millstone Grit, consisting of sandstones, fire-clays, thin coal-seams, ironstones and limestones. This is the group of strata known in England as the "farewell rock," because below it no coal-seams are found. In Scotland, however, as we have shown, the conditions are different. About 5% of the Lanarkshire coal is obtained from seams below the Millstone Grit, which are worked by about 14 collieries. The remaining 95% of coal is got from the true Coal Measures above the Millstone Grit, worked by about 250 collieries. The coarse sandstones of this group have been much used for the making of millstones, from which fact the name was derived. In Lanarkshire the fire-clays are the most valuable deposits of the series. They are worked in the northern part of the basin at Glenboig, Gartcosh, and Garnkirk. They are clays eminently suitable for the making of bricks that must withstand the action of fire. The alkaline compounds found in ordinary clays are absent, so that the fire-clays are highly infusible. They are generally supposed to have been the soils of that far-off time when the Millstone Grit was being deposited, and the absence of alkalies is ascribed to the extraction of these compounds by plants, but recent investigation has shown that all the occurrences cannot be explained in this way.

The Coal Measures occupy a large area in the centre of the county. They stretch from Glasgow up the Clyde to Carluke and Stonehouse, and extend eastwards by Coatbridge and Airdrie right across the county boundary. Most of the area lies north of the Clyde, and nearly all the large industrial centres are situated north of the river on this formation. The rocks consist of sandstones, shales, marls, fire-clays, coal-seams and ironstones. There are eleven coal-seams, of which the most important are, in descending order, the Ell, the Pyotshaw, the Main, the Splint, the Virtuewell, and the Kiltongue. Some of the seams rest on a bed of fire-clay, representing the old land-surface, while in other cases the under-clay is absent. In the former case the coal has probably been formed on the actual spot on which the forests grew,

Fossil Trees at Whiteinch

while in the latter case the vegetation may have been drifted to its present position. At Whiteinch, near Glasgow, can still be seen the remains of an ancient Carboniferous forest. The boles and roots of several fine trees have been exposed, and this unique "fossil grove" is now carefully protected in the interests of science. The bands of ironstone vary in number in different parts of the basin from four to seven. The constant repetition of sandstones, shales, ironstones and coals throughout these strata suggests that land conditions prevailed, alternating with periods of comparatively slight submergence.

The Carboniferous beds are pierced in many places by dykes and sills of igneous rocks. The dykes occur in wall-like masses, and the sills as great horizontal sheets of rock. Where the latter are thrust along a coal-seam the coal is totally destroyed and a field may be rendered unworkable. Well-known examples occur in the neighbourhood of Shotts, near Carluke, and to the east of Glasgow. The prominent ridge on which Glasgow Necropolis is situated is formed by one of these sills.

Throughout the whole extent of Lanarkshire the solid rocks are in large measure hidden by deposits of gravel, sand, and thick sheets of tough clay studded with boulders. In parts of the Clyde basin these deposits are over 300 feet thick. They project the mind back to a time when the climate of Scotland was very different from what it is to-day, when the sites of the present cornfields and orchards were occupied by glaciers creeping down from their gathering grounds, the great ice-fields lying among the high ground to the north and to the south. The stones in the boulder-clay show that two main streams of ice met in Clydesdale, the one from the Highlands, the other from the Southern Uplands. The opposing ice-sheets were then deflected both to the east and to the west, one part moving to the North Sea, the other to the Firth of Clyde. It is the ground-moraine of these ice-sheets that now forms the boulder-clay of

Lanarkshire. Gilmorehill and Garnethill in Glasgow are merely huge accumulations of boulder-clay left by the ice.

There are deposits of the glacial epoch known as "kames" that are better developed near Carstairs and Carnwath in Lanarkshire than in any other part of Scotland. Beyond the low, flat stretches of peat and moss about the Mouse Water there suddenly rises a tumbled sea of ridges and little peaks, beautifully green in colour and smooth of outline, and forming a remarkable contrast to the black peat-hags in front. The ridges are composed of sand and gravel, and wind about so as to enclose little lakes of clear water or little basins of peat marking the sites of former tarns. According to some writers the kames have been caused by denudation out of glacial débris, but it seems far more probable that they have been deposited by water in the shapes they now have against the front of the retreating ice-sheet.

It is only in parts of Lanarkshire that the soil is favourable for agriculture. In the upper ward the soil on the whole is poor and thin, and tilled land is scarce. In the centre and west of the county the ground is cold and clayey, with tracts of bog and peat. Where the volcanic rocks occur tillage is in general impossible, for the soil forms a mere film on the surface of the hard rock. Even in the lower ward the soil was originally bleak and mossy, although now vastly improved by care and cultivation. The most fertile part of the shire lies along the Clyde and its larger tributaries. Here the soil is a rich alluvium brought down by the river. Even in the upper reaches the flat, alluvial haughs are green, fertile and wooded, and contrast strongly with the bare and treeless slopes on either side.

7. Natural History.

Many centuries ago the British Isles formed a part of the continent of Europe. Where the waters of the English Channel and the North Sea now ebb and flow, there was dry land offering a free passage to the migration of plants and animals from Central Europe to our country. Such was the case when the palaeolithic hunters, the men who chased the mammoth and the reindeer with their rude stone weapons, lived in Britain. By neolithic times, however, when our primitive ancestors were using finely chipped and polished weapons of stone, the British Isles had become separated from the Continent, and Ireland was severed from Great Britain. The land-bridge existed after the disappearance of the great ice-sheet from this country, and plants and animals from Europe migrated to Britain. The land connection, however, did not remain long enough for all the continental forms of life to find their way to Britain, for we find that there are fewer species in Great Britain than in Western Europe, and fewer species in Ireland than in Great Britain. The comparative poverty of animal species in Britain is most marked in the case of the mammals and the reptiles, since these do not possess the power of flight. Thus while Germany has about 90 species of land mammals, Britain has only about 40. There is not a single species of mammals, reptiles, or amphibians found in Britain that is not found on the Continent, and only one bird, the common Red Grouse of Scotland, does not occur in continental Europe.

The plants of Lanarkshire are fairly representative of the whole of Scotland. There is, however, no mountain of sufficient height to exhibit well the peculiar Alpine plants of Scotland, although these are found in the lower basin of the Clyde outside Lanarkshire.

The moors of upper Clydesdale afford typical examples of the flora of the Scotch grouse moors. The old Caledonian forest probably existed over many areas that are now bare of trees. There is a charter extant giving to the inhabitants of Crawford parish permission to cut wood in the Forest of Glengonar, where there are now only two or three solitary trees. The existing woods of Lanarkshire have practically all been planted by man. Deciduous trees are best developed in the locality of the Falls of Clyde. Some of the individual trees of Lanarkshire are magnificent specimens. At Lee Castle there is an oak nearly 24 feet in girth, in the hollow trunk of which it is said that Cromwell and a party of his followers dined. The "Covenanters' Oak" at Dalziel House is 19 feet in girth, and there are two giants in Cadzow Forest over 21 feet in girth. There is a beech at Daldowie 111 feet high, and a poplar at Mauldslie Castle 119 feet in height.

The uplands of Clydesdale away from the river are mainly moor and marsh. In autumn they are purple with the flowers of the ling (*Calluna vulgaris*) and the fine-leaved heath (*Erica cinerea*). The milk-wort, the bog-asphodel, and in wetter parts the cotton-grass are abundant. In the marshes also the butter-wort and the sundew set their traps for unwary insects. All summer the grassy uplands are bright with the tiny, yellow flowers of the tormentil and the beautiful mountain-pansy. In damper parts are found the stone-crop and the golden saxifrage, the cinquefoil, the bog-bean, and the beautiful grass of Parnassus.

The hedge-rows are not nearly so rich as in the more genial climate south of the border, though this fact is due partly also to the general stiffness of the soil, which is as a rule derived from boulder-clay. The locality of the Falls of Clyde however is especially rich. The wood-vetch and the rarer wood-bitter-vetch are found, and here also the crane's-bill and the rock-rose are abundant, though uncommon in other parts of Lanarkshire. At the falls can be found the rare narrow-leaved bitter-cress and the purple saxifrage. The cowslip as a rule is rare in Lanarkshire, but near Bothwell it is abundant. In this locality, too, can be found the bird-nest orchid and the dusky crane's-bill. Below Bothwell Brig, round Kenmuir and Carmyle, the plant-lover is often seen, and here he may find the large loosestrife, the great leopard's bane, and the goat's beard. Possil Marsh is a favourite resort for Glasgow botanists, and several species are commonly found here that are rare in any other part of Scotland.

The mammals of Lanarkshire are quite typical of Scotland as a whole. Only three species of bats are known, the long-eared bat, the common bat, and Daubenton's bat, of which the last is distinctly the least common. The hedgehog and the mole are everywhere abundant. The common shrew is plentiful, the small shrew, though rare, has been found in the south of the county, and the water shrew has been seen near Glasgow. The wild-cat is now extinct, and the badger has also practically disappeared, although it has been seen in recent years near Carluke and at Milton Lockhart. The pole-cat and the pine marten are extinct, but the fox, the stoat, the weasel, and in suitable places the otter, are fairly common. Most of the British rodents occur in Lanarkshire, but the dormouse and the harvest mouse are not found. Of rats and mice, the old black rat is extinct, but its supplanter the brown rat is everywhere. Ubiquitous also are the house mouse and the field mouse, the field vole and the water vole are abundant, and the black vole has been recorded. The squirrel is also fairly common. Rabbits are so abundant as to become pests, and the common hare and the mountain hare are fairly often seen.

The white cattle of Cadzow Forest are in some respects the most interesting animals in the county. They are pure white in colour, except the muzzle, the hoofs, the ears, and the tips of the horns, which are black. They are popularly supposed to be the direct descendants of the ancient wild cattle of this country, the mighty *Bos primigenius* or *Urus*, but most naturalists consider them to be derived from an ancient stock of domesticated cattle. The great antiquity and interest of the breed are however undoubted, and they are the only specimens to

White Cattle in Cadzow Forest, Hamilton

be found in Scotland. The red deer is extinct in Lanarkshire, but the roe deer and the fallow deer are found.

Many parts of Lanarkshire offer very favourable opportunities for the study of bird life. In the higher parts the number of species is comparatively small, but along the well-wooded banks of the Clyde and its tributaries the variety and abundance of the birds are remarkable. A list of all the birds of Lanarkshire cannot be attempted here, but a few of the more typical species may be referred to. Of the birds of prey, the sparrow-hawk is still not uncommon in wooded districts. The kestrel is still fairly plentiful, sometimes appearing in the heart of Glasgow; but the merlin, formerly by no means rare, has now been persecuted almost to extinction. The barn owl, the tawny owl, and the short-eared owl are fairly often seen. On the quiet reaches of the river the beautiful kingfisher can still be found. The song thrush and the blackbird make the spring melodious in almost every part of the county. The missel thrush, the redwing, and the fieldfare are everywhere abundant. The chiffchaff is rare in Lanarkshire, though frequent in other places. The redstart, stonechat, wheatear, and sedge warbler all make their appearance in early summer, and the garden warbler and the grasshopper warbler are also common. The great tit, the long-tailed tit, the blue tit, and the cole tit are all abundant, while the marsh tit, though very rare in some counties, is common in the orchard district of the Clyde. The pied wagtail and the white wagtail are common generally, but the yellow wagtail, though frequently seen in lower Clydesdale, is rare in the upper part of the shire. The great grey shrike is a regular winter visitor, and there is a doubtful record of the red-backed shrike having nested in Lanarkshire.

The greenfinch, the goldfinch, the chaffinch, and the bullfinch are all found. The linnet, the crossbill, and the lesser redpole are not uncommon, and the rare mealy redpole has been seen. In the towns the starling and the house sparrow abound everywhere, and the very rare tree sparrow is recorded from Carmichael parish. Rooks and crows can be seen in all parts, but the hooded crow is rare, and the chough and the jay are now extinct in this county. Other rare birds that have been recorded are the nightjar, the wryneck, the roller, the rose starling, the snow bunting, and the waxwing. The swift, the swallow, the house martin, the sand martin, the cuckoo, and the skylark are of course abundant everywhere.

Of the goose family the teal, the widgeon, and the tufted duck are common. Of the doves the wood pigeon is common, and the stock dove is known to nest in the county. Black grouse and red grouse are common. The capercailzie was quite extinct about the end of the eighteenth century, but the descendants of introduced species have extended to Lanarkshire. Other birds important to the sportsman, the pheasant, the partridge, the snipe, the moor-hen, all are common. On the moors the mournful cry of the lapwing and the curlew can everywhere be heard.

Compared with the Continent the reptiles and amphibians of Britain are remarkably few in number. Most of the British species are found in Lanarkshire. There are two species of lizards, the lizard proper (*Lacerta vivipara*) and the blind-worm or slow-worm. The former may often be seen on a hot day frequenting dry, sunny places such as stone-heaps, walls, or ruined buildings. The latter is common among dead wood, decayed leaves, or stone-heaps, generally preferring a dry situation. The slow-worm is of course not a snake as is often supposed. It is a timid, inoffensive and perfectly harmless creature. When caught it becomes so rigid through fear that it easily breaks in two. It is from this fact that its specific name "fragilis" is derived. Of the true snakes there are two species, the adder or viper and the smooth or ringed snake. Although the latter is very uncommon in Scotland, it has been seen in the woods near Carluke. The adder is the only poisonous reptile in the country. To the healthy adult its bite is practically never fatal, although death has resulted in the case of children and infirm persons. The adder loves dry, warm places, among ruins or under fallen trees or on sunny banks. It is not common, but can hardly be called rare. The frog and the common toad abound, but the natterjack toad has never been recorded. The common newt and the palmated newt are everywhere abundant.

8. Weather and Climate.

The weather of Britain depends largely on the distribution of atmospheric pressure over these islands. To put the matter in its simplest form, when the barometer is high we expect good weather, and when the barometer is low we expect wet and stormy weather. These two types of weather correspond respectively to a condition of high atmospheric pressure or anticyclone and a state of low atmospheric pressure or cyclone. The winds in a cyclone are often strong and swirl round the centre of lowest pressure in great spirals with a direction opposite to that of the hands of a clock. When anticyclonic conditions prevail, the winds are light and move round the area of highest pressure in the same direction as the hands of a clock.

Generally speaking we may say that the winds of Scotland throughout the year are controlled by three fairly permanent pressure centres. There is a low pressure area south of Iceland, an Atlantic high pressure area about the Azores, and a continental area in eastern Europe and west Asia that is high in winter time and low in summer time. In winter as a rule the Icelandic and the continental centres predominate, as they are then working in harmony. The tendency of both centres is to draw the air in a great swirl between them from south-west to north-east. Thus we find that in winter, south-west winds predominate in Scotland. (See p. 47.) Occasionally the continental anticyclone spreads as far as Scotland, and then for a few days in winter we experience clear skies, keen frosts, and very light winds. All too soon the Icelandic cyclone centre reasserts itself, and we are back again to storms of sleet or rain with a higher temperature. In summer the Atlantic high-pressure centre has more influence. It tends to draw the winds more to the west, sometimes to north-west. This high-pressure area with its accompanying fine weather is now at its most northerly limit, and occasionally spreads over these islands, reaching the south of England frequently, but not so often extending to Scotland.

This shift of the prevailing winds from south-west to west according to season can be shown very plainly by reference to Lanarkshire records. We shall take the records of Glasgow Observatory[1], both because they are typical of the county and also because they can be absolutely relied on, which, as we shall see later, is not the case with all the records of the shire. Instead of giving numerical tables, the results are expressed as diagrams from which the prevailing winds may be seen at a glance. Along each of the eight principal points of the compass we mark a distance proportional to the percentage of days on which the wind blew from that direction, and so get a star the longest points of which show the winds that blew most frequently. The top figure on p. 47 shows that the winds during January are chiefly from the south-west, and the second figure shows that the winds during July are chiefly from the west. Easterly winds are commonest in late spring and early summer. In May they are more frequent than winds from any other direction.

The prevailing winds throughout the whole year can be shown in the same way. The third figure on p. 47

[1] For most of the data in this chapter referring to Glasgow the writer is indebted to Professor Becker of the University Observatory, who generously gave full access to the manuscripts containing the valuable meteorological records of the Observatory.

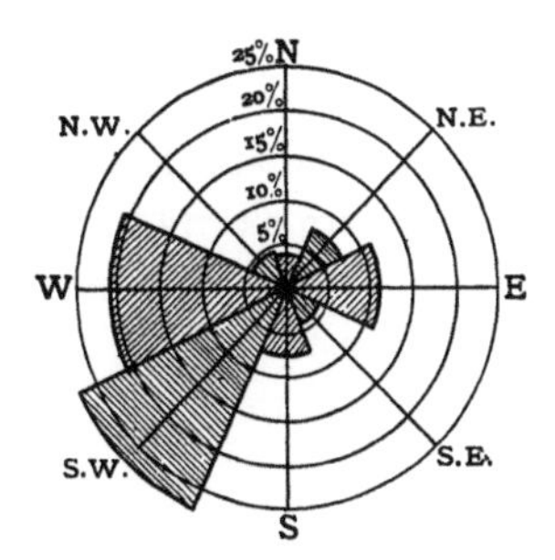

Wind Rose showing the prevalent winds at Glasgow in January

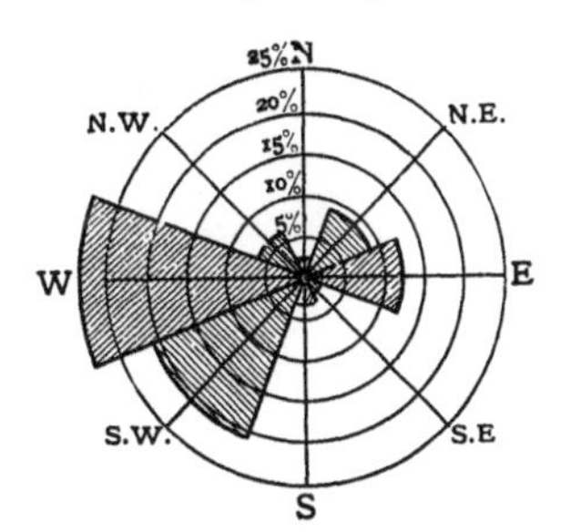

Wind Rose showing the prevalent winds at Glasgow in July

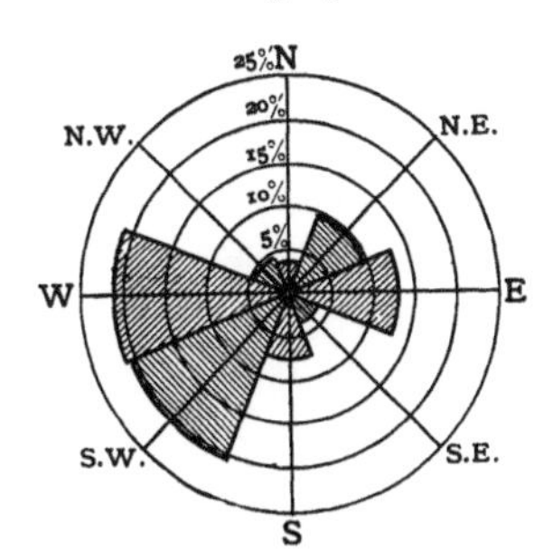

Wind Rose showing the prevalent winds at Glasgow throughout the year

shows the directions of the wind at Glasgow for the year. West and south-west winds are clearly the most common. The three wind stars show average conditions for the 16 years, 1893–1908. In many parts of the country the trees are inarticulate witnesses to the same fact. They grow with their branches pointing east or north-east, away

Tree showing S.W. wind

from the wind. The branches of the tree shown in the photograph point almost exactly north-east.

It is a general belief in this country that storms are more frequent and violent at the time of the equinoxes than at any other time. The phrase "equinoctial gales" is heard so frequently that the assumption it implies is accepted without question. It is an interesting point, therefore, to consider if the phrase is truthful. Examination of actual records proves that the so-called equinoctial gales are mythical. Storms are not more frequent at the equinoxes than at any other time. This has been clearly shown in America, where the myth is also well-established, but the records for Glasgow during the last 40 years are quite convincing on the point. They show that storms are most frequent in winter and least frequent in summer. The maximum number occurs in January, and the number decreases steadily till June and July, then rises steadily again to January. The actual figures are very interesting and are as follows:—

Number of gales over 40 miles per hour at Glasgow for 40 years, 1868–1907:

Jan.	Feb.	Mar.	Ap.	May	Je.	Jy.	Au.	Sep.	Oct.	Nov.	Dec.
50	42	36	11	5	2	2	5	10	15	27	39

During these 40 years there were four storms that for a period of 15 minutes attained a velocity of 76 miles per hour, that is, the violence of a hurricane. Not one of these was at an equinox and only one in the same month as an equinox. The dates of these four record storms were (1) January 24–25, 1868, (2) March 12, 1871, (3) October 20–21, 1874, (4) January 6, 1882.

The prevailing south-west winds of this country in winter have much to do with our favourable winter climate. The climate of the British Isles in winter is milder than that of any other part of the world in the same latitude. The following comparison will illustrate this very strikingly. Aberdeen and Nain (Labrador) are in the same latitude. The mean temperature of the coldest month at Aberdeen is 35° F., or *three degrees above the freezing-point.* The mean temperature of the coldest month at Nain is – 4° F., that is *thirty-six degrees below freezing-point.* Most of us learned at school that our good fortune as regards climate was due to the beneficent influence of the Gulf Stream, but in recent years this explanation has been entirely abandoned. It is a myth as fanciful as the supposed "equinoctial gales." The Gulf Stream becomes a negligible factor a little to the east of the Newfoundland banks. Our true benefactor is the wind. In winter time the south-west winds blow from the warm, southern regions of the Atlantic, raising the temperature of Britain and depositing moisture, which means a still further rise owing to the liberation of the latent heat. In addition they blow the warm surface waters of the ocean from more southerly latitudes and cause them to flow round and past our islands. There is no strongly-marked current, but a general "Atlantic drift" of the heated surface waters.

The temperature conditions of Lanarkshire are similar to those of other counties on the western slope of Scotland. The summers are cooler and the winters are milder than on the east coast. The mean temperature in Glasgow for January, taking an average of 40 years, is 38·6° F., and the mean temperature for July is 57·5° F., giving an annual range of 19° F. The annual range for Edinburgh is 21° F., and for London is 26° F. As the height of the land above sea-level increases, the temperature becomes lower. Thus the mean temperature for the year at Glasgow Observatory, 180 feet above sea-level, is 47° F.; at Baillieston, about 200 feet above sea-level, it is 46·7° F.; at Carnwath, a little less than 700 feet above sea-level, it is 45° F.; and at Douglas Castle, nearly 800 feet above sea-level, it is 44·8° F.

Lanarkshire is not so favoured in the way of sunshine as many other parts of the country. As a rule the amount of sunshine can be judged from the rainfall. Districts with a high rainfall have little sunshine, and conversely a low rainfall means much sunshine. The amount of sunshine diminishes as we go from south to north or from east to west. The average number of hours of sunshine per annum at Glasgow is 1095, while on Ben Nevis the amount is less than three-fourths of this figure, namely 735 hours. Aberdeen, on the other hand, has 1401 hours of sunshine per annum. The temperature and the sunshine are important factors in crop-raising. For example, wheat needs a hot, bright summer to ripen properly, and therefore we find that Lanarkshire is not an important wheat-growing county. In proportion to its size Fifeshire grows eight times the amount of wheat that Lanarkshire does.

Since in our country the moist winds come from the west, we find that the eastern counties of Scotland are distinctly drier than the western. Of even more importance is the effect of altitude on rainfall. The greater the altitude the heavier the rainfall, and if a rainfall map of the country be compared with an orographical map (that is, one showing increasing height above sea-level by a different tint of colour), the resemblance is very striking. The marked effect on rainfall of increase in altitude is

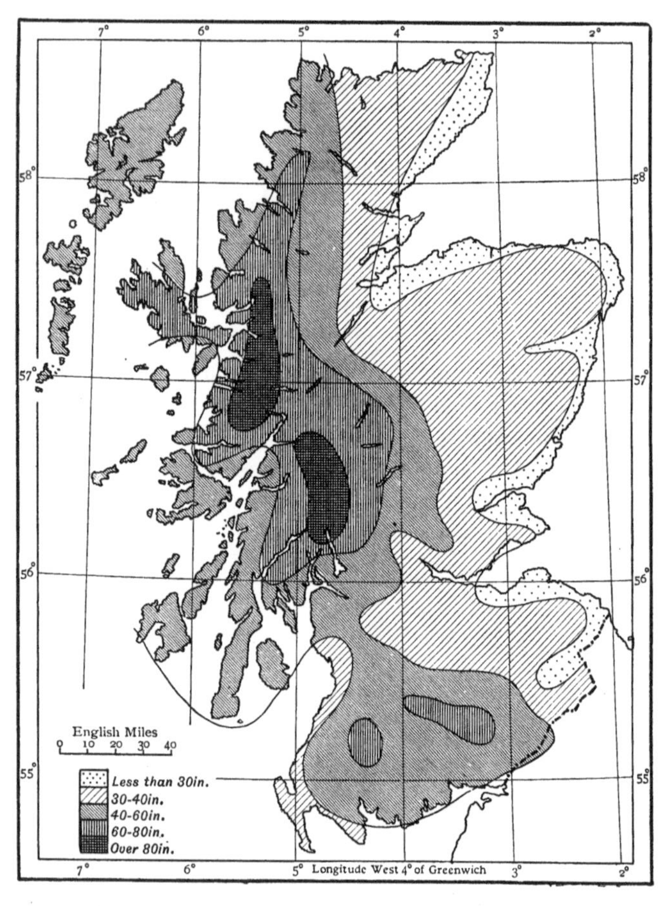

Rainfall Map of Scotland. (After Dr H. R. Mill)

shown by the fact that on the summit of Ben Nevis the average rainfall is about 160 inches per annum, while at the foot of the mountain in Fort William the annual rainfall is 73 inches.

The records of rainfall for Lanarkshire are neither so numerous nor so reliable as could be wished. Many of the earlier records must be looked on with considerable suspicion. In the New Statistical Account of 1845 an average of 30 years' rainfall at Glasgow is given as 22·3 inches, and the maximum during that period as 28·55 inches. More recent observations, however, taken over periods of 10 to 25 years show an average rainfall of 38 or 39 inches, and a maximum of well over 50 inches. Either the rainfall of Lanarkshire has altered to an amazing extent or the early records are untrustworthy, and the latter is the likelier explanation. At Dalserf, again, the average rainfall "drawn up from the observations of a medical gentleman" is given as 21·7739 inches. The "medical gentleman" who calculated his rainfall to the ten-thousandth part of an inch (!) should almost certainly have added over 50 per cent. to his figures. No wonder the worthy clergyman who gives these figures remarks with pardonable complacency, "These results, if compared with those in places lying considerably to the east, will be found to be in favour of this part of Scotland."

Remembering that the rainfall increases with height above sea-level and also to a less extent as we move to the west, we can understand the distribution of rain in Lanarkshire. The highest rainfall occurs near the head waters of the Clyde, the mean annual amount at Leadhills being over 60 inches. As we descend the river the amount gradually diminishes, until when we reach the orchard region it is little more than half that amount. Lower down the change in level is insignificant, and is

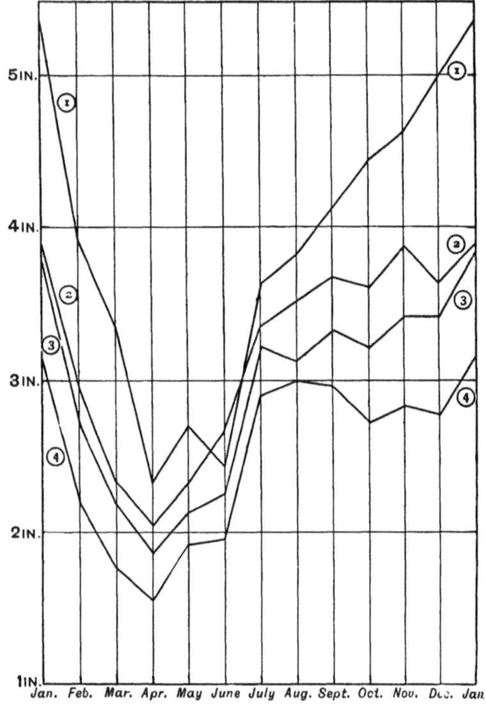

Curves showing the rainfall throughout the year at (1) Douglas Castle, (2) Glasgow, (3) Cambuslang, (4) Blantyre

more than counterbalanced by the approach to the west coast. The rainfall, therefore, increases again to about 40 inches where the Clyde leaves the county below Glasgow.

In Lanarkshire as a rule the driest month of the year

is April, and the wettest is January. This can be clearly seen from the curves on p. 54, which show how the rainfall varies from month to month. The curves show the average rainfall for each month of the year at four places in Lanarkshire, and have been drawn from the mean values of 25 years' rainfall. Although the total amounts for the year are quite different, yet the fluctuations from month to month show a similarity that is astonishing, particularly in the curves of 2, 3, and 4. Early spring is much the driest season of the year, and winter is the wettest. The curves show very clearly the sudden rise in the rainfall that takes place in July and August, a phenomenon that is but too well known to holiday-makers in the west. As regards length of daylight, dryness and hours of bright sunshine, June is undoubtedly our ideal month of summer.

The following table, compiled from the data given in the annual volumes of *British Rainfall*, shows the average rainfall over the ten years 1899 to 1908 of several selected stations in Lanarkshire:

Airdrie	38·1 inches.
Biggar	30·1 ,,
Bothwell	33·9 ,,
Carluke	34·8 ,,
Cleghorn	38·7 ,,
Glasgow Observatory ...	38·8 ,,
Hamilton	37·5 ,,
Lanark	31·5 ,,
Leadhills	63·5 ,,
Motherwell	32·7 ,,

Results for the four years 1905–1908 are given for Lamington and Slamannan as 41·6 inches and 37·2 inches respectively. But this period was rather below the average, so that if we make allowance for that fact the corrected figures will be approximately 43 inches and 38·7 inches. Buchan gives results for the 25 years 1866 to 1890 for the following places:

Blantyre	...	...	29·74 inches.
Cambuslang	...	...	34·67 "
Douglas Castle...		...	45·68 "

9. The People—Race, Language, Population.

The earliest inhabitants of Britain probably crossed from the continent of Europe when it was connected to these islands by a land-bridge. They used very roughly made stone weapons and were mighty hunters, chasing the reindeer, the mammoth, the wild-horse and other animals that lived in this country in those days. From their stone weapons they are called palaeolithic (ancient stone), and their nearest representatives in modern times are believed to be the Bushmen of Africa. Authorities are almost unanimous in maintaining that there is no evidence that this race reached Scotland. These early palaeolithic men were followed by a race that used stone weapons of a much finer type (neolithic), and relics of this race are found all over Scotland. Stone implements of neolithic type have been frequently found in Lanarkshire, and will be referred to again in the chapter on Antiquities.

One of the most constant and valuable physical characters of a race is the shape of the skull, which may be classed as long or broad. The primitive race of Scotland were long-skulled, short in stature and probably very dark in complexion. They are known as Iberians, and have no affinities with Celts or Teutons, who are of Aryan stock. Later on Scotland was invaded by Celtic tribes, who were broad-skulled, and who are generally supposed to have driven out or exterminated the Iberian race, for in early historic times the language of almost the whole of Scotland was Celtic with, however, a number of non-Aryan peculiarities of syntax. Yet it is a remarkable fact that the majority of the people in Scotland at the present time are long-skulled. Now the Teutons are long-skulled; but we know from history that the Scottish Highlanders are not of Teutonic stock, and in addition the Teutons are fair while the Celtic-speaking races are very much darker in complexion than the people of other districts. It would seem therefore that the Celtic invaders were merely a predominating and ruling caste, who completely imposed their language on the conquered tribes but did not seriously dilute their blood. The aboriginal stock absorbed the invaders, and thus on the whole the inhabitants of Scotland may be said to be of pre-Celtic or of Teutonic blood. No definite agreement on these points, however, has yet been reached.

The earliest records relating to the Clyde valley state

that it was in possession of the Damnonii, a Celtic-speaking tribe. At the end of the fourth century, when the Roman legions were withdrawn, Clydesdale was invaded by the Scots, a Goidelic tribe, and the original inhabitants were driven to the south of the district. About the beginning of the fifth century the Teutonic race began to appear in Scotland, and for 500 years this immigration went on until practically the whole of the Lowlands was in the hands of Teutonic tribes, the ancestors of the present Lowland Scots.

The place-names of Lanarkshire are extremely interesting. They are not nearly so exclusively Celtic as in the districts bordering the firth. The names of the hills illustrate this. The Celtic *bens*, *stobs*, *sgurrs*, *maols* and *mealls* are as a rule conspicuous by their absence, although we meet with the Celtic words *dun*, *torr* and *cairn*. We find the Anglo-Saxon *laws*, *dods*, *hills* and *rigs*. In the names of the rivers, however, we meet with Celtic words chiefly, such as *Clyde*, *Avon*, *Douglas* and *Calder*. The names of villages and towns fall into two classes. Those, the history of which stretch furthest back such as *Lanark*, *Glasgow*, *Dunsyre*, have generally Celtic names, while those founded in more recent times as *Roberton*, *Motherwell*, *Lamington*, have names of English origin. Several words of Norse origin also occur, such as *fell* and *gill*, and town names as *Biggar*, *Busby* and *Bearholm*. In fact, as Sir Herbert Maxwell tells us in his *Scottish Land Names*, "There is perhaps no district in Scotland where the intermixture of languages is so perplexing as in the southern part of Strathclyde."

At the present time Lanarkshire contains between one-fourth and one-third of the total population of the country. The census of 1901 gave 1,339,289 persons to Lanarkshire out of 4,472,043 for all Scotland. No

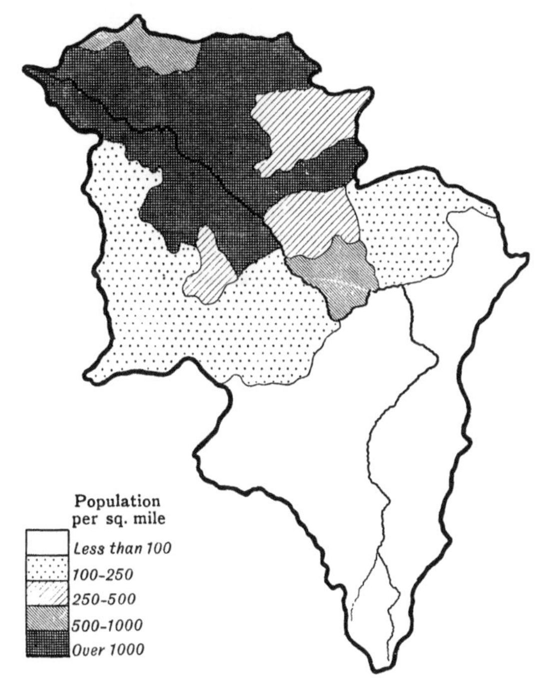

Map showing density of population in Lanarkshire

other county even approaches this number, Edinburghshire coming second with less than half a million. Although much of Lanarkshire is bare moorland, it is yet the most densely populated of all the counties, having

1524 persons to the square mile. This contrasts very markedly with Sutherland, which has only 11 to the square mile, or even with Scotland as a whole, which has 150 persons to the square mile. (See Fig. 3, p. 167.) It is only in the last hundred years that Lanarkshire has shot to the front so conspicuously, and this has been due to the industrial development of the county, following on the exploitation of its rich coal and iron fields. Motherwell has now a population of over 30,000, yet at the beginning of the nineteenth century it did not exist even as a village. In 1801 the population of the county was 147,692, that is to say it has increased during the nineteenth century almost ten-fold, while in the same time the population of the whole country has only tripled itself. The curves on p. 61 show the comparative growth during the nineteenth century of Lanark and of Edinburgh, the county next to it in importance.

The alien element is stronger in Lanarkshire than in any other county of Scotland. Every nationality into which foreigners are grouped by the census authorities is represented in Lanarkshire, and this doubtful distinction is shared by no other shire in the country. At the time of last census (1901) there were three foreigners in Sutherland, three in Nairn, two in Kinross and 13,438 in Lanarkshire, more than half the foreigners in the whole of Scotland. Of the population of all Scotland less than one-half per cent. are foreigners, but in Lanarkshire the proportion rises to just over one per cent. The nations most strongly represented are Russia, Poland, Italy and Germany. The proportion of Russians and Poles is nothing less than astonishing. Nearly four-fifths of the total number of these peoples in Scotland are to be found in Lanarkshire. These figures are largely accounted for by the number of Russian Jews engaged in various occupations in Glasgow, and the continual influx of Poles,

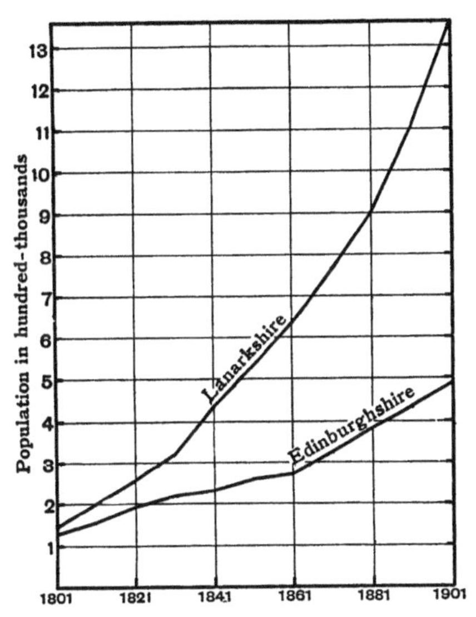

Curves showing the comparative growth of the populations of Lanarkshire and Edinburghshire

not single spies but in battalions, to Hamilton and the surrounding colliery districts. There is many a John Smith or Sandy Macgregor in these parts whose knowledge of English scarcely goes beyond his new agnomen. The Poles have the reputation of being good workers and respectable neighbours. Their occasional outbreaks on festive occasions are almost invariably confined to their own circle. There are others, not foreigners, whose knowledge of English is insignificant or absent. In 1901 there were in the county nearly 27,000 persons who spoke both Gaelic and English, and 104 who spoke Gaelic only, and of the latter, curiously enough, only eight did not reside in Glasgow or its suburbs.

Lanarkshire, in spite of its wealth and industrial supremacy, compares unfavourably with other parts of Scotland in some respects. Of its families, 71 per cent. live in houses of one or two apartments, a striking contrast to the fact that the percentage for all Scotland is 51.

The occupations of the people of the county are numerous and varied. Naturally those engaged in industrial pursuits form by far the majority of the workers. Of a total of over 400,000 men engaged in occupations of all kinds, over 300,000 are industrial workers. Those engaged in commerce total little more than a fourth of the latter number. The professions account for nearly 17,000, while agriculture claims less than half that amount. Among the industries the various branches of metal and machinery manufacturing absorb the greatest numbers, almost exactly 100,000, while next in importance come mining and quarrying, in which nearly 56,000 men are engaged.

Of course the conditions are different with women workers. Household duties for which no salary is paid are not considered "work" by the census, so that over 300,000 women are (nominally) unoccupied. Of the others nearly 100,000 are engaged in industries, while less than 38,000 are employed in some branch of domestic service. In the textile industries women take a prominent position, nearly 25,000 of them being so employed, a total that is more than double the number of male workers.

10. Agriculture.

Although at the present time Scottish gardeners and Scottish farmers have a world-wide reputation, yet it was not till the eighteenth century that there was any agriculture worthy of the name in Scotland. Most of the country was unenclosed, roads and bridges were almost unknown, artificial drainage was not employed, and only the driest parts were tilled. Yokes of oxen dragged a rude plough far up the hill sides, because the lower parts were hopeless swamps. A few sentences may be quoted from Henry Grey Graham's description of the state of agriculture at the beginning of the eighteenth century. "There were no enclosures, neither dyke nor hedge between fields, or even between farms; so that when harvest began or the cereals were young, the cattle either required to be tethered, or the whole cattle of the various tenants were tended by herds."..."When the harvest was over the cattle wandered over all the place, till the land became a dirty, dreary common; the whole ground being saturated with the water which stood in the holes made by their hoofs. The horses and oxen being fed in winter on straw or boiled chaff, were so weak and emaciated that when yoked to the plough in spring they helplessly fell into bogs and furrows; even although to fit them more thoroughly for their work, they had been first copiously bled by a 'skilful hand.'"..."The harrows, made entirely of wood,—'more fit,' as Lord Kames said, 'to raise laughter than to raise soil,'—had been in some districts dragged by the tails of the horses, until the barbarous practice was condemned by the privy council." ..."If one man dared to cultivate a neglected bit of ground, the others denounced him for infringing on their right of grazing on the outfields. How could he begin the growing of any new crop? The others viewing every innovation with the contempt which comes from that feeling of superiority, which ignorance and stupidity produce, would refuse to join him."..."With a system so atrocious, with land uncleaned, unlimed, unmanured, undrained, it frequently happened that the yield could not feed the inhabitants of the district, and men renting from 40 to 100 acres needed to buy meal for their families."

Gradually new crops and better methods were introduced. The cultivation of turnips and potatoes marked the beginning of a more rational agriculture. Old ideas, as the determination to use no mechanical aids to winnowing because it contravened the Scriptures and "was making Devil's wind," gradually disappeared. Stock-breeding was introduced, the land was let in larger holdings, alternation of crops was practised, artificial fertilisers were used, until at the beginning of the nineteenth century agriculture was on a satisfactory basis.

In some respects Scotland will always be at a disadvantage compared with England. In many parts the soil is as fertile as any south of the border, but the more favourable climate of England causes an earlier harvest. An additional crop of turnips or cabbages or vetches can then often be secured after the main crop has been got in, whereas in Scotland this can very seldom be done.

On the whole Lanarkshire cannot be considered one of the chief agricultural counties of Scotland. Much of the ground, particularly in the upper ward, is quite unsuited for tillage. Thus it follows that only about a quarter of the total area of the county consists of cultivated land, whereas in Fife and Haddington more than one-half of the land is cultivated. The only branch of agriculture in which Lanarkshire excels is that of fruit-growing. Its orchards have been famous for many centuries. From Lanark to Bothwell both banks of the Clyde are devoted to fruit-growing. Apples, pears and plums of the finest quality have been grown here since the beginning of Scottish history. Gooseberries and currants claim a fair proportion of the area, but in recent years strawberry cultivation has increased enormously. Even Perthshire, the other great fruit-district of Scotland, has less than half the acreage of Lanarkshire devoted to this fruit. The former county excels in raspberries, however, which are grown only to a moderate extent in Lanark.

A common practice is to grow strawberries for three or four years. The land is then ploughed and a corn crop is taken. Next year potatoes are grown for the purpose of cleaning and enriching the soil, when the ground is heavily manured and is ready again for strawberries. Over four tons of strawberries per acre can be gathered from the fields bordering the river. Higher up the banks, the yield is not so heavy, but the fruit is considered of finer quality. The tomato is quite the newest incomer to the district, but its cultivation has spread with remarkable rapidity. Everywhere the little glass houses are springing up like mushrooms to meet the rapidly increasing demand for this fruit in the large towns of the district.

Of the area under orchards proper, most is claimed by plum trees, and in autumn the wayfarer may see for miles along the road within easy reach the scarlet fruit gleaming through its green setting. Apple trees are almost as common, and both these species are grown over twice the area given to pears. Cherry cultivation is relatively unimportant. The total area under different kinds of small fruit in 1908 was 2259 acres, and under orchards was 765 acres.

Scotland is not a great wheat-growing country; the summers are too wet and cold. In fact, in several of the counties not a single acre of land is given to wheat. By far the most important crop is oats, which is peculiarly well suited to our climate. In Lanarkshire, for example, oats occupy more than twenty times the area devoted to wheat. This contrasts very markedly with some of the English counties such as Cambridge, where wheat is grown over nearly twice the extent occupied by oats. The comparison with a purely agricultural county like

Cambridge is instructive. The latter county is not quite so large as Lanarkshire, but it has nearly 52,000 acres of oats to 37,000 in Lanark, and actually 93,000 acres of wheat to 1700 in Lanark. There are no other corn crops of any importance in the shire, but of other products turnips and potatoes are the most valuable. There are over 9000 acres under turnips and 5000 under potatoes. The area given up to hay is, of course, extremely large, there being nearly 48,000 acres thus cultivated in 1908. (See Fig. 4, p. 168.)

In stock-raising also Lanarkshire cannot compete with many of the other counties of Scotland. Compare it with Aberdeenshire, for example, and the contrast is striking. The latter contains over 31,000 horses used for agricultural purposes, while Lanarkshire can boast of barely 8800. It is of interest to note, however, that one of the best known breeds of horses in the world is of Lanarkshire extraction. It is said that the famous Clydesdales originated from the crossing of a Flemish stallion with a Scotch mare in the seventeenth, or as some say in the eighteenth century. They were certainly brought to a high pitch of perfection in the upper ward of Lanarkshire during the eighteenth century. The type is not unlike the mighty English "shire" horse, but, to quote an authority, "the English breed is larger, and possesses more substance than the Clydesdale, but the latter has a decided superiority in bone and muscle, with a compact and firmly knit body, symmetrical head, and strong feet and pasterns, that render its strength more durable and admirably fit it for heavy draught work."

Lanarkshire ranks third among the counties for cattle, being surpassed only by Aberdeen and Ayr. The total in 1908 was more than eight times the number of horses, namely 71,636. The cattle are kept chiefly for dairy purposes, and therefore the great majority of them are Ayrshires, although a number of Highland cattle may be seen in the upper ward. The former breed has been

Clydesdale Stallion

found peculiarly suitable to the moist climate of the south-western counties. It is not only hardy, but yields a larger proportion of milk to food consumed than any other breed in the country. Glasgow and other large towns absorb the supply of most of the dairies, but cheese is made in some parts, particularly round Carnwath and Lesmahagow.

In the upper ward the green hill slopes form fine pasture grounds for sheep, and there are many large sheep-farms in the district. The stock consists chiefly of Cheviots and Black-faced sheep. The wool of the Black-face does not bring so high a price as that of the Cheviot, but the former breed is hardier and more suited to mountainous tracts. It will thrive on poor fare and withstand privations that would exterminate any other breed. In 1908 the number of sheep in Lanarkshire was 257,779. (See Fig. 5, p. 168.)

The ancient Caledonian forest probably at one time extended over most of Lanarkshire, but only a few doubtful vestiges of this now remain in Scotland. In certain parts of Lanarkshire there were undoubtedly within the last thousand years forests that have now entirely disappeared. At the present time the woodlands in the locality of the Falls of Clyde show a better development of deciduous trees than any other part of the west of Scotland. Over 21,000 acres of the county may be classed as woodland, and within the last ten years this amount of land under trees has just barely held its own.

11. Industries and Manufactures.

There is probably no district in Britain where the variety of industries and manufactures is greater than in Lanarkshire. This is, in truth, one of the most significant features of the county. Specialisation, to any marked extent, is absent. We do not find groups of towns engaged almost exclusively in the cotton trade as in Lancashire, or in the woollen trade as in West Yorkshire, or in the iron trade as in the "Black Country" of England. This is undoubtedly a favourable state of affairs, for it is seldom that several of the great industries are notably depressed at the same time, and sudden fluctuations from excessive prosperity to the depths of adversity are not nearly so common in Lanarkshire as in other great manufacturing districts.

The pre-eminence of Lanarkshire as an industrial centre is due to several causes, of which the most important is the possession of valuable coal-fields of large extent and fine quality. The position of these coal-fields must not be overlooked. In the chapter on Geology they were shown to stretch as far down the Clyde as Glasgow, and thus the manufactures of Lanarkshire have ready access to the markets of the world. It was about the middle of the eighteenth century before the manufactures of the county began to develop. The stimulus given to the textile trades by the application of machinery in the second half of the eighteenth century was felt in Scotland. The inventions of Hargreaves, Arkwright, Crompton and Cartwright in England practically laid the foundations of a new industry, and Lanarkshire was not slow to seize its opportunities. Other industries followed, reacting one upon the other, until the county became a hive of varied industries—the spinning of cotton, silk and flax, weaving and dyeing, the production of pig-iron, the rolling of steel, the firing of pottery, glass-making, the building of bridges, the manufacture of chemicals, distilling and brewing, and a thousand and one other industries from the building of battle-ships to the making of clay tobacco-pipes.

For over a century the production of iron has been a leading industry of Lanarkshire. For many years the West of Scotland was the most important iron district in Britain and now ranks second only to Cleveland. The

Ironworks, Coatbridge

first blast-furnaces in the county were begun at Wilsontown (Carnwath) in 1781. In 1788 there were only eight blast-furnaces in all Scotland, turning out less than 7000 tons of pig-iron in a year. At the present time there are nearly 60 in Lanarkshire alone. Coatbridge is the chief centre for this branch of the iron trade, more than half the blast-furnaces of the shire being situated there. The neighbourhood of Glasgow is next in importance.

The two most marked advances in the production of pig-iron have both originated in Lanarkshire. At first charcoal was used as fuel, and later coke, but in 1831 Messrs Dixon introduced the use of coal, thus effecting an enormous saving. The consumption of coal per ton of iron produced has fallen from eight tons to less than two tons. The other striking innovation was the introduction of the hot blast, suggested by James B. Neilson in 1828. Experiments were made at the Clyde Ironworks with complete success, and soon every furnace in the country adopted the idea.

The production of mild steel, which began about 1872, led to the demand for purer ores than could be found in the district. This necessitated a great import of iron ore, chiefly from Bilbao in Spain, in addition to supplies from England, Algeria and Elba. Steel-making is now one of the most important industries in Lanarkshire. The Steel Company of Scotland was the pioneer firm and was founded in 1871. It owns works at Newton and Glasgow, and uses the Siemens or open-hearth process. There is, it may be noted, no Bessemer steel made in Lanarkshire. Motherwell is now recognised as the centre of the Scottish steel industry. The Dalzell Steel and Iron Works (Colville's) in this town have the largest plant in Scotland and can turn out 5000 tons per week. On every side in Motherwell indications of the predominant industry assail one's eyes and ears. The air resounds with the clatter and bang of the rolling-mills, the clanging of the steam-hammers and the rattling fusillades of the pneumatic riveters. At Parkhead Forge near Glasgow, armour plates are the speciality, and this firm can now build a battleship, protect it with armour plates, fit it with boilers and even supply it with guns. This last fact is particularly interesting, for it marks a new industry in Lanarkshire. The first modern gun from this county was completed towards the end of 1909. It is a 12-inch gun firing a projectile over a third of a ton in weight, and nothing like it has ever been made in Scotland before.

To describe in detail the multitudinous industries based on iron and steel would need many volumes and only a rapid glance can be given at them in these pages. Foundries are numerous, producing castings which vary in size from the parts of a model engine to the gigantic cylinders of a battleship. Boilers of all kinds, Lancashire, water-tube and ordinary marine types, are made in Glasgow, Motherwell, Govan and other places. Many works, again, devote themselves to machine tools, half-human contrivances for punching and shearing, for rolling and bending, for planing and sawing. Machinery of every kind, in fact, is manufactured in the industrial towns of the county-land and marine engines, cranes, pumps, steam-hammers, winding-engines, sugar-machinery and innumerable other kinds.

The making of scientific instruments has an added interest from the long connection of Lord Kelvin with this branch of industry. The name "Kelvin and White" is known wherever Glasgow-built ships go. It was the invention of the mirror galvanometer by Lord Kelvin that made possible communication across the Atlantic.

First modern gun made in Scotland

Over 50 patents were taken out by him, and with many pieces of scientific apparatus Glasgow practically supplies the world. Another Glasgow invention of great interest and importance is the range-finder of Barr and Stroud. Their design was adopted by the Admiralty, and is now fitted on all battleships and cruisers. It is used also in nearly every other navy in the world.

The building of locomotives has been brought to a high pitch of perfection in Lanarkshire. The Hyde Park Locomotive Works, Springburn, Glasgow, in many respects are the premier works not only of Britain but of all Europe. Three hundred engines in a year can be turned out, and when one thinks of the wonderful complexity of the modern locomotive this is an astonishing figure. The north-east of Glasgow is in fact devoted to locomotive building, for there also the Caledonian and North British Railways have their works. Other locomotive firms too there are in the city; the trade employs many thousands of men, and Glasgow engines can be seen in every quarter of the globe.

Roof and bridge work is carried on in various parts of Lanarkshire, chiefly in Glasgow and Motherwell. Those stupendous examples of human achievement, the Forth Bridge and the Tay Bridge, might justly be regarded as among the wonders of the world. They were constructed by the well-known Glasgow firm of Sir William Arrol & Co., and there are other builders in the county of hardly less eminence.

The Clyde and ship-building are synonymous. The first passenger steamer ever launched in Britain was built

Latest type of Locomotive (Atlantic type). Hyde Park Works

on its banks, and at the present time it is the greatest ship-building centre in the world. In 1907 the tonnage built on the Clyde was nearly double that produced by the whole of Germany. Every kind of sailing craft that can be called a ship will be found a-building here, from a racing yacht to a Lusitania, from a square-rigged wind-jammer to a battle-ship. It is only the north-western extremity of Lanarkshire, from Glasgow seawards, that takes part in this industry. The Renfrew and Dumbarton banks are lined with yards of the first importance, turning out a tonnage far exceeding that of Lanarkshire, but it must not be forgotten that their existence depends largely on this last county. Their coal and steel come from Lanarkshire, and in several cases the workers themselves travel from Glasgow and its suburbs to the ship-yards and back again every day.

The most famous yard in Lanarkshire is that of the Fairfield Company, Govan. They have built several of the most famous Cunarders, but it is in warships that they take the most prominent position. Up to the end of 1908 they had built for the British navy ships in total displacement amounting to almost 200,000 tons, a figure equalled by no other firm on the Clyde.

The textile industries of Lanarkshire, although important, are not to be compared with those of Yorkshire or Lancashire, where whole communities devote themselves to nothing else. The weaving industry, however, is one of the oldest in the county and in the sixteenth century was firmly established. After the Treaty of Union, when new markets were opened to Scottish enterprise, the trade grew rapidly, and by the beginning of the nineteenth century, spinning and weaving provided employment not only in the towns and villages of Lanarkshire but in lonely cottages far from urban districts. But in the struggle for existence the factory operative conquered the hand-loom weaver, who, in many instances, had either to emigrate or starve. The industry is now concentrated in large factories.

Cotton-spinning is largely carried on in Glasgow and the surrounding district. The east end of the city also still produces muslins and curtains on quite a large scale. The subsidiary industries of bleaching, dyeing and printing employ large numbers of people. The manufacture of linen used to be an important industry in the west of Scotland, but the competition of Belfast has been too keen, and now the trade is practically extinct. Worsteds and woollen cloth goods are made to some extent, but the most important branch of the woollen industry in Lanarkshire is the manufacture of carpets. Glasgow carpets have a very great reputation for high quality, and some of the best designers in the country were for the first time induced to enlist their talent in the service of Templeton of Glasgow. The manufacture of silk fabrics, though not increasing to any marked extent, still holds its own. Handkerchiefs, ties, chiffons and other light materials are the chief articles produced.

The chemical manufactures of Lanarkshire are characterised, like the industries in general, by great variety. Perhaps the most important product is sulphuric acid, or oil of vitriol, which is produced in large quantities by several makers. Other acids, bleaching powder, and Epsom salts are also important products. In recent years the increased appreciation of the value of fertilisers has given a stimulus to the manufacture of artificial manures. Even ironworks are engaged in this business, for the spare gases from the furnaces are not allowed to escape, but are made to yield their share of ammonia from which to make ammonium sulphate, a valuable fertiliser.

It is estimated that twenty years ago there were only fifty tons of potassium cyanide per annum consumed in the whole world. The invention of the cyanide process for gold extraction entirely altered these conditions, and created an enormous demand for cyanide. The process was evolved in a Glasgow laboratory by MacArthur and Forrest, and the patents were in the hands of a Glasgow firm. The demand for cyanide at the present time may be imagined from the fact, that the use of it for a period of five years on the Rand alone has recovered thirty-five millions sterling of gold. The Cassel Gold Extracting Company of Glasgow have a manufacturing capacity more than double that of any other works in the world.

A very interesting industry that has sprung up in Glasgow during the last few years is the making of rubber-cored golf balls. Hundreds of girls are now employed in this manufacture, and the industry is growing rapidly. There are many other branches of chemical industries that can only be mentioned. Among them the most important are sugar refining, brewing, distilling, the manufacture of oxygen, the making of paints and varnishes, soap making, oil-distillation, tanning, starch and gum making, electro-plating, and waterproofing.

Pottery and glass-making have for long been staple industries in Lanarkshire, particularly in Glasgow. Many different kinds of glass are made, and the products have a wide reputation. In recent years, however, the competition of France and Germany has been more severely felt, perhaps because the local manufacturers are not so modern in their methods as their continental rivals. Of the various branches of glass-making, probably the one that has been brought to the highest degree of perfection in Lanarkshire is the manufacture of globes and shades for gas and electric lights.

The clay for use in the potteries is obtained largely in the south of England, and is brought to Glasgow in small sailing vessels. It varies in quality and also in the ingredients added to it according to the class of ware desired. The clay used always to be worked on the potter's wheel, an instrument that was in use five thousand years ago, and is only now becoming extinct owing to the introduction of machinery.

The presence of valuable beds of fire-clay in different parts of Lanarkshire has resulted in a very important and flourishing industry in the west of Scotland, namely, the making of fire-bricks, retorts, pipes, troughs, garden vases, and many other articles. Ordinary building bricks are also made in large quantities, and the supply could be largely increased, but the presence of so much good building stone in Lanarkshire limits the demand for brickwork. In recent years an interesting method of using waste material in brick-making has been discovered. This consists in the utilisation of the great "bings" or heaps of blaes that form too conspicuous a feature in the landscape of many parts of Lanarkshire. The material is crushed and then moulded into bricks under high pressure.

12. Mines and Minerals.

Lanarkshire has been an important mining centre for many centuries. Although there is no definite information on the point, it seems very likely that the lead mines of the upper ward were worked by the Romans. From many points of view, however, the most interesting mineral found in Lanarkshire is gold. The history of the gold workings makes a fascinating story. The early Celtic tribes of the district certainly made torques and other ornaments of gold, specimens of which have been found in different parts of the county. The gold of which these ornaments were made must almost certainly have come from upper Clydesdale. When we come to the beginning of the sixteenth century in the reign of James IV, we find that the mines of Crawford were well known, and several valuable finds are recorded. A manuscript in the British Museum tells us that there were three hundred miners at work in this reign. The writer states shrewdly "that there hath ben...plentie of golde gotten in ye waters of the said cloughes and Gilles 80 fad[oms] above the foresaid waters in ye valleis, wch golde being ponderous...must bie common reason descend: so as consequentlie, whereas some peeces of [gold] of above 30 ounces weight have been found in the said Gillies, the same must...growe there aboute or bie violent waters be dryven out of higher places wher they did grow within ye circumference of those places where the golde is founde."

Tradition tells us that in the reign of James V the French ambassadors were hunting with the king near Crawford. They taunted the king with the poorness of his country till, stung by their jeers, James wagered that the district produced richer fruit than any in the fair land of France. His wager was won when, at the banquet that evening, instead of fruit the ambassadors were served with covered dishes containing "Bonnet pieces," coins made of the gold found in the neighbourhood.

In 1542 crowns both for the king and the queen were made of gold from the Leadhills district, and these can still be seen in Edinburgh Castle. During the minority of James VI, a Dutchman, Cornelius de Voss, formed a company and prosecuted the search with such vigour that in thirty days gold worth £450 was sent to the mint at Edinburgh. Another company, headed by a Fleming, was not so successful, and James ended their license. A number of good reasons for so doing was given, but the best undoubtedly was, as the act states, "and which is most inconvenient of all, has made no sufficient payment of the duty to our Sovereign Lord's treasury." Later on our Sovereign Lord James conceived a very characteristic and ingenious "plot" to make the mines productive. He suggested to Bevis Bulmer, a mining expert of the day, that twenty-four gentlemen should each contribute £300, and the king would make each a knight, "a Knight of the Golden Mynes or a Golden Knight."

This Bevis Bulmer,

> "Who won much wealth and mickle honour
> On Shortcleuch Water and Glengonar,"

was the most famous of the gold miners of Lanarkshire. Working with a staff of 300 men he secured in three years gold to the value of £100,000 sterling. It is very interesting to note that he erected a stamping mill at the head of Longcleuch Burn, for he had found a "little string or vein powdered with small gold." Many attempts have since been made to find the gold *in situ* but without success. All the gold obtained is found among the stream gravels and clays. Bulmer's friend and pupil, Stephen Atkinson, tells us that he had "too many prodigall wasters hanging on every shoulder of him...and at last he died in my debt £340 starling, to my great hindrance: God forgive us all our sinnes."

Atkinson obtained power to continue the work and to make "ane new searche, tryall and discouerie of the mynes, seames and minerallis in Crawfurde Mure," but his elaborate project was not successful. Throughout the three centuries since that time, gold has been collected in small quantities from the Leadhills and Wanlockhead district. Little nuggets have occasionally been found as large as a bean, but most of the stream washings of gold are in the form of fine grains. The miners still turn out on special occasions, such as the marriage of one of the

Hopetoun family, and obtain enough gold to make the wedding-ring. There are mining experts who are of the opinion that with the application of modern, economical methods, gold-mining in Lanarkshire might be made a commercial success.

The lead mines of Lanarkshire have certainly been worked for nearly seven centuries. In a grant of lands to the monks of Newbattle in 1239 by Sir David Lindsay, a lead mine on Glengonnar Water is mentioned, and in 1264 the sum of forty-two shillings is entered in the accounts of the sheriff of Lanarkshire for the conveyance of lead from Crawford to Rutherglen. Lead-mining in the old days was a more exciting occupation than it is now. In spite of guards the wild Borderers occasionally raided the lead-bearers, and even certain staid burgesses of Lanark and Glasgow were accused of seizing a quantity of lead on its way to Leith, and were ordered to restore their stolen goods. With varying success the lead-mining was carried on until recent times. In 1810 about 1400 tons of lead were produced, but towards the middle of the century the output diminished to about seven or eight hundred tons. The mines were then taken over by the Leadhills Mining Company, and soon the industry was placed on a prosperous footing, so that by 1892 the output of dressed ore amounted to nearly 2000 tons. The plant, however, was old and out of date, but during the last ten years the company have embarked on the bold policy of adopting modern methods and putting down expensive machinery with complete success.

As this is the only district in Scotland where lead is

mined at the present day, a few details may be given regarding the methods of working. Hauling, pumping and lighting are now partially done by electricity, and the system is being extended. To this day the ore is entirely hand-mined, but compressors are being put down for rock drilling. When the ore (galena or lead sulphide) has reached the surface it is first hand-picked, the lumps of pure ore being thus extracted. This is almost all exported to India. The residue is washed and crushed, and the rock is separated from the ore by gravitation. The ore is mechanically graded according to size, and is sold in the condition of pure, dressed ore. Until two or three years ago the ore was smelted on the spot, but owing to complaints of farmers regarding the injurious effects of the lead fumes on vegetation, the smelting was discontinued. Even the washing water is not allowed to escape without paying its toll of lead. It is run into circular troughs and set rotating. The lighter sand, owing to centrifugal force, settles round the outside of the tank, while a heavy lead mud is recovered from the centre. In 1908 the quantity of ore produced was 3199 tons, while the neighbouring mines at Wanlockhead just over the Dumfriesshire border produced less than half that amount.

By far the most important mineral in Lanarkshire is coal. It forms the foundation on which the whole industrial success of the county is based. The method of occurrence of the seams has already been described in the chapter dealing with the geology of the county. The two methods of extracting the coal are known as the "stoop and room" system and the "long wall" system, and both methods are largely used in Lanarkshire. In the first method roads are driven through the coal and connected by cross-passages, leaving pillars of coal to support the roof. The roof is afterwards propped up by timber and the coal-pillars removed. This method is generally employed for thick seams. For thin seams the long wall system is preferred. As the work proceeds outwards the whole of the coal is extracted, and the "face" is thus gradually pushed outwards, while the waste material is stacked up to support the roof. In recent years in Lanarkshire coal-cutting machinery has been largely introduced. It is used on the long wall system for thin seams, and is often made to cut through the under-clay, thus preventing any waste of coal. About two-thirds of the machines are driven by electricity and the rest by compressed air.

Lanarkshire is the most important county for coal in Scotland. In 1908 it produced over seventeen million tons, the next county being Fife, with exactly half that amount. The coal is not all used locally, as there is a considerable export trade from the Clyde. There are more than 55,000 people employed at the coal mines of Lanarkshire, and this number and the coal-output increase from year to year. It is therefore a point of very great importance to know how long this enormous drain on the coal resources of the county can last. It has been estimated that there are probably between one and two thousand million tons of coal left in the ground. Even taking the higher figure and assuming that the production will not increase, it is plain that the coal of the county will be exhausted in little more than a century. Almost certainly the easily got and therefore cheap coal will be

Houses cracked by underground workings, Motherwell

exhausted before then, and it is cheap coal that makes Lanarkshire the great industrial centre that it is.

In various parts of Lanarkshire the extraction of minerals, particularly coal, has had various effects on buildings. Some towns exhibit every appearance of having been visited by an earthquake. Gaping cracks run through the walls of some of the houses, others are supported by beams and stays, and others have become so dangerous that they have had to be deserted altogether. In Motherwell quite a number of the houses show how their foundations are gradually giving way, and builders and purchasers of new property have to be extremely particular regarding their choice of a site.

In the early days of the iron industry only local ores were used. A great impetus to the mining of iron ore was given in 1801 by Mushet, who discovered that the miners were rejecting under the name of "wild coal" the valuable ore known as blackband ironstone. For many years there was no need to import foreign ore, but the advantages of foreign haematite for steel-making and the gradual exhaustion of the better seams of ironstone have caused a great change in this respect. The output of Scottish ores has fallen off rapidly. Thus in 1881 the production of Lanarkshire and Ayrshire was 2,232,237 tons, in 1890 it was 721,793 tons, and in 1908 it was 432,840 tons, not a fifth part of what it was thirty years ago. In the production of ironstone in 1908 Lanarkshire ranked second among the counties, being beaten by Ayr, which had nearly 270,000 tons.

The extraction of fire-clay is now an important industry in Lanarkshire. Mining is carried on chiefly at Glenboig, Garnkirk, and Gartcosh. The fire-clay occurs in beds from four to twenty feet thick, and is of a very high quality. In 1908, 354,000 tons of fire-clay were mined in Lanarkshire out of a total for the whole country of 880,000. Large quantities of ordinary clay are also dug for the making of building-bricks. Over 200,000 tons were extracted in 1908.

The oil-shales of Scotland are found chiefly in the east of the country, Linlithgowshire being by far the largest producer. In the eastern part of Lanarkshire, however, a considerable amount of oil-shale is mined; the quantity in 1908 being nearly 45,000 tons. The oil-shale industry is a peculiarly Scottish one. It was in 1850 that James Young made the important discovery that paraffin oil and solid paraffin could be obtained by the distillation of certain shales. A flourishing industry sprang up, and Scotch oil was exported to every part of the globe. In recent years the history of the industry has been one of continual struggle against the competition of the enormously rich oil fields of America and Trans-Caucasia. The industry still flourishes only through the far-seeing policy of applying technical skill of the highest kind to the different manufacturing processes, resulting in improved methods and therefore diminished working expenses.

In connection with the mining industry it seems strange to remember that little more than a hundred years ago slavery existed in Scotland. Many of the coal-hewers and coal-bearers were serfs, compelled to labour all their lives in bondage. A workman who dared to leave his pit was in the eyes of the law a thief, for he had stolen himself from his master. If his children once went to work in the mines they were slaves thenceforth. Even children in their infancy were sometimes sold by needy parents to the coal masters. This monstrous state of affairs was only partially remedied in 1775, and it was not till 1799 that tardy justice gave unconditional freedom to all.

13. Shipping and Trade.

Glasgow is the gateway through which enters and leaves the mighty double stream of trade that continually pours in and out of Lanarkshire. Moored to its ten miles of quays lie ships that have come from every corner of the globe—here a sailing-ship laden with ore, battered and rent by its long voyage from the far South Seas; a hundred yards away a floating hotel plying across the Atlantic with the regularity of a river-ferry. The shipping that now enters and clears from Glasgow each year amounts to over five million tons.

The history of the shipping trade is a continuous record of the triumph of human determination and forethought over opposing natural forces. The originator of Glasgow's commerce overseas is said to have been William Elphinstone, who about 1420 exported salmon and herrings to France, and brought back brandy and salt in exchange. By the middle of the seventeenth century most of the inhabitants were engaged in commerce, and traded with Ireland, the Western Islands, France and Norway. By the end of the century Glasgow's mercantile marine numbered 15 ships, having an average register of nearly 80 tons. It was not, however, till after the union of the Parliaments that Glasgow's trade showed a rapid growth. The ports of England and the English colonies were now thrown open to Scotland, and soon Glasgow's colonial trade was of considerable importance. Ships were at first chartered from other ports to bring back the tobacco from

The Broomielaw

Virginia, but in 1718 the first Glasgow-owned vessel, a Greenock-built ship of 60 tons, crossed the Atlantic. The trade was so successful that several English ports formed a "combine" against Glasgow, and complained to government regarding the fraudulent dealings of the Scottish merchants. Investigations ensued, resulting in the acquittal of the Glasgow merchants without a stain on their characters, the finding being that "the complaints are groundless and proceed from a spirit of envy."

During the latter half of the eighteenth century more than half the tobacco imported into the kingdom was brought to Glasgow, and made the fortunes of the "tobacco lords," who strutted in their scarlet cloaks on the "plain-stanes" of the Trongate, ignoring the appealing looks of the mere shopkeepers, who, when they wished to do business, stood in the gutter meekly awaiting an opportunity of catching the eyes of the great men. The outbreak of the American war in 1775 dealt a crushing blow to the tobacco trade, from which it never fully recovered, but Glasgow enterprise and Glasgow capital soon poured into other channels, and a flourishing trade with the West Indies arose.

The introduction of steam-navigation marked the beginning of a new era in the history of shipping, and the first successful steamboat may justly be considered to be Symington's "Charlotte Dundas," which was built about 1801 and plied for a short time on the Forth and Clyde canal. Ten years later Henry Bell's "Comet" was built, and soon was followed by several other steamers. About the same time a new trade was opened up with the East Indies and proved so successful that in a very short time it reached large dimensions.

The most famous line of steamships in the world, the Cunard Line, was founded by Messrs Burns in 1840 with the "Sirius," and this was the first steamship to cross the Atlantic. Soon afterwards the hardly less famous "Anchor" and "Allan" lines were formed, at first for trade with America only, but soon sending ships to many other countries. At the present time there is regular communication with almost every port in the British Isles, all the great ports of the Mediterranean and Western Europe, with Canada and the United States, with South America, India, China, Japan, the West Indies, and Australia.

In 1907 the total tonnage that entered and cleared at Glasgow was over five millions, a tonnage nearly double that of Leith, which ranks as the second port of Scotland. The export and import trade in the same year was valued at over £46,000,000. In spite of a few fluctuations, the trade of Glasgow still seems to be growing rapidly in value, as in 1895 it was only about half the 1907 value. Glasgow generally ranks fourth among the ports of Great Britain, the other three that surpass it in value of trade being London, Liverpool and Hull. It is somewhat surprising, however, to find that in 1907 it had to yield pride of place to Manchester, a practical illustration of the value of a ship-canal.

The imports are chiefly food-stuffs, namely wheat and flour, animals and meat. The wheat and flour come principally from the United States, the Argentine Republic, India, Canada, Russia, and Australia. We receive our supplies of cattle and sheep chiefly from the United States and Canada, and our dead meat from the United States, the Argentine Republic, Denmark, New Zealand and Canada. The exports, as might be expected, consist almost entirely of manufactured goods, the most important being machinery and iron and steel goods.

Cotton goods rank next in importance, and coal, linens and spirits are valuable items. Glasgow has always formed one of the chief emigration ports of the country, particularly for the United States and Canada. Most of the emigrants are natives of Scotland, but many are foreigners who cross to this country from the Continent, generally in large batches under supervision, in order to obtain the advantage of Glasgow vessels.

The growth of Glasgow has been accompanied by the relative decline of Rutherglen. It was a royal burgh in the twelfth century, and for centuries it was the chief trading town of lower Clydesdale. Even at the beginning of the fifteenth century, when Lanarkshire was divided into two wards, Rutherglen was the chief town in the lower ward. It asserted active superiority over Glasgow and levied tolls from the Glasgow inhabitants until the fifteenth century. In matters of trade, however, the people of Rutherglen were not so far-seeing as their canny Glasgow neighbours, who consistently clung to their commercial ideals. Thus in the sixteenth century, when Glasgow, Renfrew and Dumbarton combined in an attempt to improve the navigation of the Clyde, Rutherglen held aloof. As a result the town is still in no more favourable a position as a port than it was 500 years ago. Small vessels can reach the town at high tide. It must not be forgotten, however, that Rutherglen has still a ship-building yard from which vessels of fair size have been launched.

14. History of Lanarkshire.

Two thousand years ago Clydesdale was inhabited by a tribe called the Damnonii. They are usually referred to as Celts, but we have already indicated the possibility that Celtic blood may not have been nearly so prominent in Scotland as Celtic culture and speech. When the Romans invaded Scotland the route down Clydesdale was one of the easiest ways from England to the Scottish Lowlands, and therefore this part was overrun by the Romans. The great rampart built on the line of Agricola's forts passes through the north-west extremity of Lanarkshire, and is known as Antonine's Wall. There was no real colonisation of Clydesdale by the Romans. It was held by the soldiers as a military outpost, and consequently we find remains of camps and well-made roads, but not of permanent settlements.

After the departure of the Romans the district reverted to its former owners, better known as the Britons of Strathclyde, and the capital of the kingdom was Alcluyd or Dunbreatan (hill of the Britons), now known as Dumbarton. In the seventh century the district became for a time subject to the Anglian King of Northumbria, and for centuries after this time there must have been a constant influx of Anglo-Saxons and a gradual expulsion of the natives. In spite of attacks from the Norsemen, the kingdom grew in power until in the tenth century it stretched as far south as Cumberland.

After the defeat and death of Macbeth, about the

middle of the eleventh century, most of Scotland was united under Malcolm, the son of Duncan. From about this time Anglo-Saxon influence predominated in the Scottish Lowlands, and Celtic influence became subsidiary. On Malcolm's death the kingdom was again divided, until the accession of David I in 1124 finally united all Scotland into one kingdom. At the court of Henry I of England, David had become imbued with Norman ideas and culture, and therefore we find during his reign an influx of Normans into Scotland, who soon settled down in permanent residence and founded some of the most powerful families in Scotland.

It was in David's time that the county of Lanarkshire probably first became an administrative district with boundaries roughly approximating to its modern limits. It was not, however, until the War of Independence that Lanarkshire came prominently to the front. Although Sir William Wallace was not born in the county, he made his home there, and many of his best-known exploits are associated with the shire. His first serious conflict with the English in Lanark has already been described. The last phase of Wallace's career is also associated with Lanarkshire. It is said that it was in the church of Rutherglen that Sir John Menteith agreed to betray Wallace, and at Robroyston, a few miles from Glasgow, he was made prisoner by the English.

Wallace had been aided by Sir William Douglas, a member of the famous house of Douglas, powerful in Scotland before the influx of the Normans. Their ancient seat was in Douglasdale, one of the most inaccessible parts of Lanarkshire. Most famous of the Douglas line was the "Good Sir James," generous and valiant friend of King Robert the Bruce. When Sir James met his death in Spain on his way to Palestine with the heart of the Bruce, he had in his company another brave Lanarkshire knight, Sir Simon Loccard of Lee. To him was entrusted the duty of bringing back the heart in its padlocked casket to Scotland, and since that time his descendants have

Lee Castle. Home of "Lee Penny"

added to their coat-of-arms a heart and padlock, and the name of the family was changed to Lockhart. Since that time also the Douglases have carried on their shields a bloody heart and a crown.

Sir Simon Loccard returned from this campaign with the famous "Lee Penny," part of the ransom of a prisoner. It is a red, heart-shaped stone, latterly set in a shilling-piece of Edward I's reign, and for centuries was used as a healing talisman. Sir Walter Scott's novel, *The Talisman*, obtains its title from this relic, as a fanciful account of the stone forms an important incident in the story.

From the time of the Bruce the county was at peace until the ambition of the Douglas family brought upon the district the miseries of a civil war. William, eighth Earl of Douglas, took refuge abroad for a time, and on his return was slain by the king's own hand. In 1455, James II demolished Douglas Castle. Passing to Glasgow, he gathered the men of the west, returned to Lanark, and then burnt and harried all Douglasdale and Avondale. Other members of the house of Douglas who figured prominently in Scottish history are Archibald Bell the Cat and Archibald, sixth earl (of the younger branch), who was grandfather of Lord Darnley, and thus great-grandfather of James VI.

The county was torn again by civil war in the time of Queen Mary. In 1544, during the Queen's minority, the Regent James Hamilton, Earl of Arran, besieged the Earl of Lennox in Glasgow. On the surrender of the garrison they were all, with the exception of two, treacherously massacred. The most important event of this period was the Battle of Langside, on which to a large extent depended the future of the whole of Scotland. When Queen Mary escaped from Lochleven her supporters assembled in force at Hamilton. The Regent Murray was encamped at Glasgow to prevent the passage of the Clyde on the way to Dumbarton. Learning that the Queen's army would attempt the passage lower down the river, Murray moved out to Langside Hill to intercept the enemy. Here he was attacked by the Queen's forces, but completely defeated them. Mary was watching the battle from a hill near Cathcart, and on seeing the flight of her army, galloped off in terror and did not stop till she reached Sanquhar, 60 miles away.

For a hundred years peace reigned in Lanarkshire, and then the flame ot civil strife broke out anew. The county became a refuge for those who sought a respite

Monument in memory of Battle of Langside

from the persecutions of the "killing times." The unfrequented hills, the wide moorlands, and the peat bogs of Clydesdale became the haunts of the Covenanters, and many of the best-known episodes of the struggle against Episcopacy were associated with Lanarkshire. The famous Peden "the Prophet" has given his name to a stream that runs down from the Lowthers to join the Powtrail Water, for here he found shelter in a shepherd's cottage beside the burn. In Crawford, John Willison had a secret chamber constructed where the persecuted might find shelter, and here Donald Cargill found refuge for a time. It was at Covington mill in Lanarkshire that Cargill was at length taken prisoner, in 1681.

Monuments of the Covenanters are dotted all over the county, in the heart of Glasgow, and on bare hill sides far from any town. One of the most interesting may be seen in Hamilton churchyard erected to the memory of four martyrs whose rudely carved heads ornament the stone. They were Lanarkshire men who were executed at Edinburgh in 1666. Their bodies were quartered, the right hands were taken to Lanark, where they took the covenant, and their heads were exhibited on the old Tolbooth at Overnewton. The inscription on the stone is as follows:

> "Stay, passenger, take notice what thou reads;
> At Edinburgh lie our bodies, here our heads,
> Our right hands stood at Lanark; those we want
> Because with them we sware the Covenant."

The affair at Rullion Green originated in Lanarkshire and Dumfriesshire. The insurgent Presbyterians assembled at Lanark and then moved towards Edinburgh. Meanwhile General Dalziel was marching from Glasgow to Lanark when he found that his enemy had given him the slip. He came up with them on Rullion Green, on the eastern slopes of the Pentlands, and there completely routed them. Many of the prisoners that were taken experienced the tender mercies of the "boot" and the "thumbscrews."

It was in Lanarkshire in 1679 that the Covenanters obtained their most noted victory over their persecutors. The 29th May was being celebrated as a holiday, since that day was the anniversary of the restoration of Charles II. The bonfires were blazing in Rutherglen when a party of Presbyterians entered the town, quenched the fires, held a brief religious service, and entered their

Battlefield of Drumclog

protest against the forcible establishment of Episcopacy, a copy of which they nailed to the cross. Graham of Claverhouse drew out his dragoons from Glasgow to avenge this affront. In Hamilton he heard of a conventicle at Loudoun Hill, and moved in that direction. He was opposed by a fairly large body of men, though poorly armed, who were stationed at Drumclog, in the upper Avon valley near the Ayrshire border. The insurgents were skilfully drawn up on a boggy piece of ground behind a large ditch. A hot engagement ensued, the dragoons were outflanked, about 30 were killed and the remainder hurled back in disorder, Claverhouse himself escaping from the field with difficulty. This success

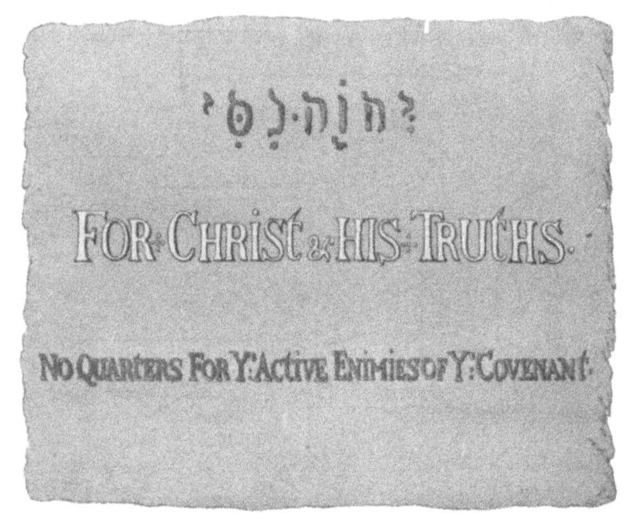

Banner of the Covenanters at Drumclog and Bothwell Brig

greatly encouraged the insurgents, and their numbers rapidly increased.

In June the Duke of Monmouth advanced against them with a powerful army. They were well posted at Bothwell Brig, across which the Duke would have to move in order to attack them, but their ranks were torn with dissensions between the moderate and the extreme parties.

Part of them behaved gallantly, but we are told that the main body seemed "neither to have had the grace to submit, the courage to fight, nor the sense to run away." Five hundred of the insurgents were slain and about 1200 prisoners were taken, many of whom were executed or sent as slaves to the plantations. A body of Scottish

Bothwell Bridge and Monument

Highlanders with Monmouth's army distinguished themselves by their cruelty.

The revolution of 1688 put an end to the persecution of the Covenanters, and active warfare was seen no more in Lanarkshire. At the rebellion of the '45, however, the road through Clydesdale was chosen by Prince Charlie in his retreat from England. The army followed the route now traversed by the Caledonian Railway from Carlisle, namely up Annandale, over Beattock Summit into Clydesdale, and then down the Clyde to Glasgow. Prince Charlie spent the last few days of 1745 in Glasgow, a most unwelcome guest. He had extorted large supplies from the city, both in money and in food and clothing, but even the compensation received by the citizens, namely, a grand review on Glasgow Green, roused no enthusiasm. He procured only 60 adherents during his stay, and these the scum of the town. In fact the provost of the time maintained that his only recruit was "ane drunken shoemaker," and it is said that but for the intercession of Cameron of Lochiel the Prince would have sacked and burnt the town.

Since the beginning of 1746 the history of the county has been one of uninterrupted progress. Its epoch-making events have been discoveries in industry, its revolutions have been revolutions of industrial methods, and the improvement in social conditions and customs has been no less marked.

15. Antiquities.

The earliest men in Britain were unacquainted with the use of metals. Their weapons and tools were of stone roughly shaped and chipped. These weapons of palaeolithic type do not occur in Scotland, but stone weapons and tools finely chipped or polished have been discovered in many parts of the country. Neolithic implements of this type have been frequently found in Lanarkshire.

Palaeolithic implement

(From Kent's Cavern, Torquay)

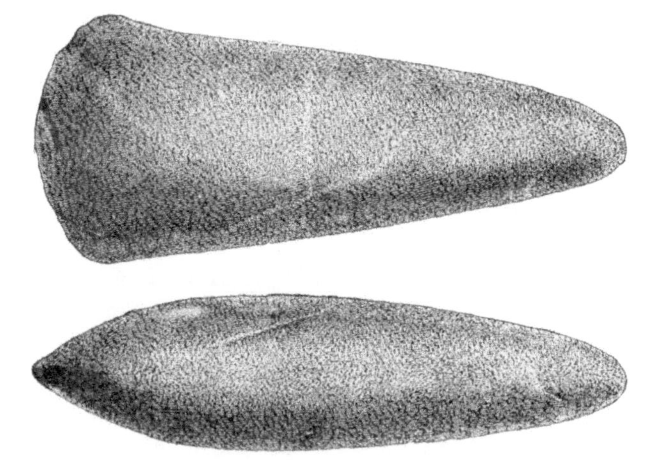

Neolithic Celt of Greenstone

(From Bridlington, Yorks.)

They consist of celts or axes, arrow-heads, spear-heads, flail-stones, knives or scrapers, slick-stones for softening hides, and other implements. Many of the so-called Druidical monuments were probably erected by Neolithic man. The race was widely distributed, stone structures of a similar kind having been found all over Europe, in Africa, Asia, and America. The cromlechs or "standing stones" of Scotland probably belong to this period. One of the finest in Lanarkshire is a megalithic pillar near Elvanfoot, the largest in the district, which gave its name to the farm on which it was found, Crooked Stone. Near Biggar, part of a circle of standing stones may be seen, and near the head-waters of both North and South Medwins similar monuments occur.

As mankind progressed in civilisation the art of metal-working was discovered, and the earliest metal implements are made of bronze. These at first imitated the shape of the stone tools, so that we find celts and other implements fashioned like the stone ones, but of bronze. Many such relics have been discovered in Lanarkshire. A fine bronze axe-head over six inches long was found near Biggar, and bronze spear-heads, celts, crowbars, and other tools have been discovered from time to time in different parts of the county.

In a British camp on the right bank of the Clyde, below the junction of Glengonnar Water, an interesting sepulchral urn was found. It was about six inches in height and made of coarse clay mixed with grit. A rough ornamentation of a herring-bone pattern had been given to it, and on being opened the urn was found to be full of calcined bones. With the urn was found a bronze armlet, which is stated to be perhaps the finest specimen of this type of personal ornament which can be met with in any collection or museum. It is nearly three inches in external diameter, nearly half an inch thick, and dates probably from the time immediately preceding the Roman invasion. Sepulchral tumuli, often containing remains of this nature, are common throughout the county. There is a tumulus in Lesmahagow parish over 50 feet high, many of the stones used in the construction of which weigh about a ton.

Ornaments of gold have been found in a few places. At Stonehill, in Carmichael parish, two rings of pure gold but of rude workmanship were dug up. Gold torques or collar ornaments have been found in more than one part of Lanarkshire. One very fine specimen was obtained near the borders of Culter parish. It is of thin gold in the shape of a crescent, the horns of which nearly meet. The middle part is almost an inch and a half broad, and the torque is decorated with lines and depressions. Another is a circular ribbon of gold spirally twisted and ending in two hooks. It is too small to have been hooked round the neck, so that if worn as a necklet the ends must have been joined by a thong, or the ornament was possibly hooked round the arm. Another specimen was found at Carmichael. They have been attributed to a time just after the withdrawal of the Roman legions. A Strathclyde poet about 650 A.D. describes how 363 warriors wearing the collar of gold went out to fight the Saxons and were all slain but three.

Several British camps have been discovered in different parts of the county. At Cairn Grife, near Crawfordjohn, is a fine example of one of these ancient British forts. The fort is about a hundred feet square, and is enclosed by two ramparts separated by a distance of from five to seven yards. Camphill in Glasgow gets its name from the remains of a British camp that can still be seen there. At Nether Abington there rises abruptly from the banks of the Clyde a hillock, partly artificial and partly natural. It is protected by a ditch on the land side, and projecting from this, a rampart and a ditch of later construction take the form of a horse-shoe. This type is common in England, but rare in Scotland. They seem to have been used first by the Saxons and then by the Normans. The mound was excavated by G. Vere Irving, who found that a fortress had been built on a sepulchral tumulus.

A crannog or pile-dwelling was discovered a dozen years ago on the margin of a pond near Hyndford. It was probably inhabited during the Roman occupation of the surrounding country, for a large number of Roman remains were obtained from it. The most interesting relic discovered was a fine torquė, consisting of large, ornamented bronze beads strung on an iron rod.

In Roman times Clydesdale was one of the chief highways from England to the Lowlands of Scotland. The chain of Agricola's forts and later the wall of Antonine ran for a short distance across the north-western part of the county, about four miles north of Glasgow. Traces of the wall can be distinctly seen to the west and to the north of Cadder. The "wall" or rampart was built of sod in layers upon a stone foundation. Its height was about 12 feet, and it was protected by a fosse or ditch nearly 40 feet wide and about 12 feet deep. A magnificent collection of inscribed stones from the rampart can be seen in the Hunterian Museum, Glasgow University. The stones include sepulchral monuments, altars to Jupiter, Hercules, Apollo, Diana, and other deities, and slabs commemorating the completion of the building of portions of the wall by the troops that set up the tablet.

From Annandale to the west end of the wall the main Roman road from England followed the Clyde. Its course was almost identical with the present route of the Caledonian Railway from Carlisle. It entered Lanarkshire by the head streams of the Annan, crossed Beattock Summit to Clydes Burn, and then followed the main stream through Carstairs and Cleghorn, past Uddingston and Glasgow to the wall. A branch road came into Lanarkshire by the Powtrail Water, joined the modern road below the Dalveen Pass and reached the main Roman road near Elvanfoot. Above Carstairs a Roman road ran at right angles to the main route. The east limb probably went into Edinburghshire, while the west limb ran parallel to Douglas Water towards Duneaton Water and Crawfordjohn. Another branch stretched through Stonehouse and Strathaven parishes to Ayrshire. In many places these roads can still be clearly made out, particularly in the upper ward of the county. They often run for considerable distances parallel to the modern road, and generally at a slightly higher level.

Along the roads a great many camps have been discovered. In the valley of the Clyde they can be counted in dozens, and they are generally accompanied by relics of some description that give clear evidence regarding their occupants. A large and important camp near Cleghorn is supposed to have been occupied by Agricola. It is rectangular in shape, its length being 600 yards and its breadth 420 yards, and there are six gates. Another important camp is situated on the right bank of Clydes Burn above its junction with the Clyde. It is not quite so large as the one just mentioned, as it measures 500 yards by 300 yards. It is believed to have been one of the temporary camps of the western column of Agricola's army. A typical Roman castellum was discovered on the right bank of the Clyde near its junction with Clydes Burn. On three sides access is difficult, and therefore a single rampart protects it on these sides. On the weak north-western side the defences were doubled. The interior was excavated and a circular basin was found chiselled out of rock, the tool marks being quite distinct. It was lined with clay for the purpose of holding water, and an abundant supply was still in the reservoir. Round Tinto there are several camps that cannot be referred to any Roman road.

Traces of Roman occupation abound in this district, and fancy can easily conjure up the picture of the conquering legions marching vigilant and irresistible through the lonely hills, where the Caledonian guerillas were waiting to take advantage of any slackness of discipline. All along the line of march we find their armour, their utensils, their ornaments, their money. At Carstairs camp were found coins of the reigns of Aurelius, Antoninus, and Trajan ; a silver coin with the head of Faustina was found at Lanark ; near Biggar many Roman coins have been obtained, one of gold bearing the head of Vespasian; and along the line of road at Burnhead and Castlehill gold coins have been discovered.

The Lesmahagow Flagon

A find made at Lesmahagow in 1810 is of particular interest. The natives had been familiar with a round stepping-stone in a certain burn. The stone became indented, and this peculiarity roused curiosity and led to its examination. It was found to be a beautiful bronze flagon with several symbolic figures, including Mercury and Minerva. It is now in the Hunterian Museum, Glasgow University. A very rare and curious relic of the middle ages was found in Culter parish. It consists of a small metal shrine about four-and-a-half inches high, and shaped like an arm. In it were kept the relics of some holy man, a portion of the arm, or perhaps only a finger. Its date is about the end of the thirteenth century.

16. Architecture—(*a*) Ecclesiastical.

The earliest *Celtic* examples of ecclesiastical architecture were dry-built stone cells with a roof closed with overlapping stones and flag-stones. These were followed by the Columban Scottish churches, consisting of one small oblong chamber with one door and one window. No ornamentation was used until the Romanesque influence made itself felt, introduced by the Normans. The type was elaborated later by the addition of a chancel.

The Celtic structures were superseded by churches of *Norman style*, introduced in the twelfth century. This style is characterised chiefly by simple massive forms and semicircular arches. As a rule there is little ornament except in the doorways, the arches of which are moulded, and into which zig-zag or bird's-head ornamentation is introduced. Very few good examples of this style exist in Scotland, but parts of the cathedrals of Dunblane and Kirkwall and the abbey of Dunfermline exhibit it very well.

The round Norman arch was replaced by the pointed arch, giving the *First Pointed Style*, which reached Scotland in the thirteenth century. Fresh ornamentation was introduced showing itself in mouldings and in vigorous foliage. The windows were always pointed, narrow and lofty, and an effect of greater spaciousness, combined with lightness, was aimed at. In Scotland the style was not so pure as in England or France, as round Norman forms lingered on, especially in doorways, although the general style was altered. Glasgow Cathedral presents a fine example in its crypt and choir, and in St Kentigern's, Lanark, the style is shown particularly well by the doorway.

From the middle of the fourteenth to the middle of the fifteenth century, the *Middle Pointed or Decorated Style* prevailed in Scotland. The details aimed at a still lighter effect. The windows were enlarged, the tracery became more ornate, and the vaulting and buttresses were made lighter. Perhaps the finest example of the style in Scotland is Melrose Abbey. The nave of Glasgow Cathedral is also a good fourteenth century example of the Decorated style.

The transition to the *Third Pointed Style* was gradual. In England the tracery became more rigid, and the windows were carried up in straight lines so that the style was called *Perpendicular*. In Scotland the exterior is generally marked by rather heavy buttresses, terminating in small pinnacles. The semicircular arch is often used, and there is a revival of early ornamentation. Most of the examples are not cathedrals but collegiate churches.

A fragment of St Mary's Church, Rutherglen, is of Norman architecture. The tower of this church, which still stands, is unique in Scotland. It had no connection with the original church, being built later. The church was granted to the Abbey of Paisley by William the Lion in the twelfth century, and remained a possession of the abbot till the Reformation. In Lamington Church, founded by Lambin in the twelfth century, a fine Norman doorway is still preserved. The elaboration of the doorway is surprising for such a remote place. There are three orders of mouldings all showing characteristic Norman ornamentation. The bell of the church bears the date 1647, and down to 1828 the church still kept its "jougs" for the punishment of evil-doers.

Old Church Tower, Rutherglen

St Kentigern's Church, Lanark, is a fine example of the First Pointed Style. It certainly existed in the twelfth century, and was given by David I to the Abbey of Dryburgh. It possessed the somewhat unusual feature of a double chamber, being divided down the centre by a row of arches. In the south wall is a fine example of a First Pointed doorway, with characteristic foliage and bold, pointed mouldings above. The church continued to be used till after the Reformation, but by the middle of the seventeenth century it had fallen into a ruinous condition.

Glasgow Cathedral was built in parts at different times, and therefore shows different styles of architecture. The earliest part of the present building dates from the

Glasgow Cathedral

twelfth century. It is a mere fragment, but enough remains to show that it was in the Transitional style in use in the second half of the twelfth century. The choir, after being destroyed by fire, was completed by Bishop Joceline in 1197, and of this building a considerable part still remains. The present choir was built in the thirteenth century by Bishop William de Bondington, and illustrates not only the genius of the architect but the wealth of the community that erected it. The mouldings are very elaborate, and the whole structure is of singular richness and beauty of design. Later on in the century the nave was built, a work distinguished by great simplicity and dignity. The mouldings are characteristic, but not so elaborate as those in the choir. The whole structure was probably roofed, and the basement of the central tower erected, by the middle of the fourteenth century. In the fifteenth century Bishop Cameron erected the stone spire of the cathedral, the details of which are especially fine.

The general form of the building, like that of all cathedrals, is a cross, but the transepts project so slightly that the long stretch of the walls seems almost unbroken. The general impression given by the exterior is that of simplicity bordering almost on bareness, but the interior is magnificent. The proportions are noble and harmonious, and the details are particularly beautiful and rich.

In 1560 the government ordered the destruction of the altars, images and other monuments of the old faith, and this barbarous edict was consequently carried out. In 1574 the Assembly instigated a further act enjoining more destruction, and then ensued throughout Scotland the devastation of many of those beautiful structures of the middle ages, in most cases replaced by Protestant churches that architecturally were beneath contempt. Andrew Melville urged the Glasgow magistrates to order the destruction of the cathedral, and they at length consented. Masons and other workmen were assembled to begin the work of demolition, but the Crafts of the

Crypt, Glasgow Cathedral, showing St Mungo's Tomb

city took arms and swore that he who cast down the first stone should be buried under it, nor would they be pacified till the workmen were dispersed. Thus Glasgow Cathedral escaped untouched, almost alone of all the cathedrals of Scotland. The building remained in a dilapidated condition till 1854, when the Commissioners of Woods and Forests undertook its restoration, and under their care it was put in its present state.

St Bride's Church, Douglas, belongs to the Middle Pointed or Decorated Period. Although the church existed in the twelfth century, the present building is much later. In 1307 the old church was attacked by Sir James Douglas, and the English garrison who were in it at the time were slain. The present church was built about the end of the fourteenth century. The architecture is very simple, but the church is interesting for the monuments it contains. The oldest is ascribed to the Good Lord James, the friend of Bruce, although some authorities hold that it dates to a still earlier time. It is surrounded by a finely cut canopy. A silver case is still preserved containing all that is left of the heart of the Good Lord James. His bones also are said to have been brought home from Spain and buried here.

"The Banys haue thai with thame tane
And syne ar till thar schippes gane
....
Syne toward scotland held thar vay,
And thar ar cummyne in full gret hy.
And the banys richt honorabilly
In-till the kirk of dowglass war
Erdit, with dule and mekill car."

Another monument is to Archibald, the fifth earl, who died in 1438. The base of the monument is ornamented with sculptured foliage, which seems to have been done about the middle of the fifteenth century. A third monument, inferior in design and execution, is to James, seventh earl of Douglas and his wife. This was James the Gross, who died in 1443. Remains of the earlier

Old Church, Bothwell

church still exist in fragments of capitals of Norman design.

St Bride's Collegiate Church, Bothwell, is another example of the same style. It was founded in 1398 by Archibald, Earl of Douglas, surnamed the Grim, who was the lord of Bothwell Castle. Here the Duke of Rothesay was married to the earl's daughter in 1400.

The church is a simple oblong building divided externally by buttresses. Above the entrance doorway is a remarkable arch of elliptic form. The roof is protected by overlapping stone slabs which are carefully curved so as to throw the water away from the joints. The church contains some fine monuments and several ancient carved stones.

The Church of St Nicholas, Biggar, is a representative of the Third Pointed Period, and was founded in 1545 by one of the powerful Fleming family, then Chancellor of Scotland. Like many others of this style the doorway is surmounted by a round arch. The walls are buttressed on the outside and there is a battlemented tower. One unusual feature is that the pointed windows are set in a rectangular recess, probably due to the square Renaissance forms then being introduced.

The Reformation put an end to mediaeval ecclesiastical architecture in Scotland. A few churches were certainly erected under the influence of the Episcopalians, but the Presbyterians attempted to eliminate everything that savoured of the old forms, and to this end were content to erect buildings that had absolutely no claim to respect so far as their architecture was concerned. One interesting example of a seventeenth century spire, a type few of which now remain in Scotland, is the Tron Steeple, Glasgow, erected in 1637. It seems to be an imitation of the steeple of Glasgow Cathedral, modified, however, according to the style of the time. The wide arches at the base are modern.

In the eighteenth century there was in England a

distinct revival of the interest in architecture, and particularly in classical styles. This awakened feeling hardly stirred in Scotland till the nineteenth century. We are told that in the eighteenth century the Scottish churches "were disgraces to art and scandals to religion. They were mean, incommodious and comfortless; the earth of the graveyard often rose high above the floor of the church, so that the people required to descend several steps as to a cellar, before they got entrance by stooping into the dark, dismal, damp and hideous sanctuaries." At the beginning of the nineteenth century, however, a great change for the better began to take place. Architects made a special study of old buildings and old styles, and this combined with the rapidly increasing wealth of the country was soon reflected in many noble, ecclesiastical buildings. The great and wealthy industrial communities of Lanarkshire can now without exception boast of modern churches that will bear comparison with those of mediaeval times.

17. Architecture—(*b*) Castellated.

Bothwell Castle, situated on a high promontory on the north side of the Clyde, is perhaps the most magnificent ruin in Scotland, and is undoubtedly the finest example of the castles of the thirteenth century. It encloses a large courtyard surrounded by high walls, strengthened at the corners with towers. In places the walls are sixty feet in height and more than fifteen feet thick. The splendid donjon, which dominates the building, dates probably from the second half of the thirteenth century. The whole structure is built of red sandstone in regular courses, and the earlier parts particularly bear witness to the wonderful care and skill expended on the building. In the north-west tower there is a drawbridge of a kind unique in Scotland. It was constructed to cut off the tower from

Bothwell Castle, interior

attacks from the *inside*, and was counterpoised for easy lifting. The great hall, the chapel and other buildings were probably erected about 1400 by Archibald the Grim, Earl of Douglas. Round the red towers of the castle linger memories of some of the most famous names in Scottish history. Edward I and Edward III of England both lodged within its walls. One of its owners was

James Hepburn, Earl of Bothwell, who married Queen Mary. The present owner is the Earl of Home.

Directly opposite the castle stand the ruins of Blantyre Priory. Only two gables and a vault are left of the priory founded in the thirteenth century by Alexander II, and gifted to the monks of Jedburgh, who found at times a refuge there when the Border was troubled. It stands on a steep bank; and Wallace is said to have leapt through a window over the cliff, thus eluding his pursuers. Tradition tells also of an underground tunnel beneath the river connecting it to the castle on the other side.

The end of the thirteenth century is marked by a great change in the style of the castles of Scotland. The War of Independence completely exhausted the resources of the country, and consequently we find that large and massive buildings such as Bothwell Castle were no longer erected. Their place was taken by strong, square towers, fashioned after the model of the Norman keeps. These are specially characteristic of the fourteenth century, but continued to be built at much later dates, and from the simplicity of the design it is often difficult to determine the exact age. In the fifteenth century it became customary to build the castle round a central quadrangle or courtyard. In addition, a separate tower or keep is often found, capable of being defended although the rest of the castle should be captured.

One of the best known of fifteenth century castles is Craignethan Castle, near Crossford, the "Tillietudlem Castle" of *Old Mortality*. It has a beautiful situation above the Nethan Water, about a mile from the Clyde.

The keep of the castle is certainly older than the other portions, and is built on a plan very unusual in Scotland. At first the keep was probably the only part erected, but later the extent of the building became considerable. The castle has been held in turn by the Hamiltons, the Hays,

Craignethan Castle

and the Douglases; and Queen Mary is said to have occupied it for some time before the battle of Langside.

Another interesting castle of the fifteenth century overlooks the town of Strathaven. It is situated on a rocky hill nearly surrounded by the Powmillan Burn, a tributary of the Avon. The castle was built by the grandson of the second Duke of Albany, who in 1457 became Lord Avondale. It seems to have been oblong in plan with two towers at diagonally opposite corners, although only one round tower and fragments of the walls remain. There are large port-holes for the mounting of guns.

Covington Tower is another example of the simple keep of the fifteenth century. The walls are eleven feet thick, and the tower must originally have been a fine specimen of the massive work of the time. The estate was obtained in the fifteenth century by the Lindsays, and was held by them till the seventeenth century. Lamington Tower is traditionally associated with Wallace, for it is said that he married Marion Bradfute, the only daughter of the Laird of Lamington. The existing tower, however, is not older than the sixteenth century. Crawford Castle or Tower Lindsay is noticed in Scottish records as early as the twelfth century. It is situated near the Clyde, commanding the road up Glengonnar Water into Dumfriesshire. The castle was a favourite resort of James V as a centre for hunting. The ruins that still exist probably date from the reconstruction of the castle by the first Marquis of Douglas, about the beginning of the seventeenth century.

Douglas Castle was the scene of many of the most famous exploits in Scottish history. It had been seized by the English, but was recaptured by Sir James Douglas in 1307. Not wishing to be besieged in it by the English, he removed all the valuables, piled up the provisions and wine in the cellars, slew his prisoners and added their corpses to the heap, cast the dead horses on the mass, and threw salt over all. He then set fire to the building with its dreadful "Douglas Larder." No trace remains of the original castle, the "Castle Dangerous" of Sir Walter Scott. The solitary tower of an old castle that still exists is probably not earlier than the seventeenth century. In 1755 a fire destroyed the whole structure except the tower.

18. Architecture—(*c*) Municipal and Domestic.

The Glasgow Municipal Buildings were finished in 1889 at a cost of over half a million sterling. The style is the Italian Renaissance, and the buildings are four storeys in height with a domed tower at each corner and a central tower rising to a height of 237 feet. The windows and other parts are elaborately ornamented with columns, and figure sculpture has been used to a large extent. The result, while extremely rich, perhaps errs on the side of over-elaboration. The interior is as magnificent as the exterior. A splendid marble staircase leads from the entrance hall, and the corridors and walls are pillared and panelled with marble and alabaster. In the banqueting-hall is a fine series of panels painted by the most eminent Scottish artists and representing scenes in the history of Glasgow.

The town hall of Rutherglen is a remarkably imposing structure for the size of the town. It is late baronial in style and bears a square clock tower with turrets. The tower forms a conspicuous land-mark for many miles around.

There are many interesting dwelling-houses in Lanarkshire, but space will permit the mention of only a few especially notable either architecturally or for their historic associations. Provand's Lordship is the oldest

George Square and Municipal Buildings, Glasgow

house in Glasgow and one of the oldest in Scotland. This interesting link with the past is situated in Castle Street on the west side of Cathedral Square. It was the town manse of one of the prebendaries of the Cathedral who was Laird of Provan, and is said to have been built in the fifteenth century. It has of course suffered considerable alteration; for example, the present level of the street is five feet above the old ground floor of the house. The interior still retains many of the old features. It is said that Queen Mary stayed here when Darnley lay sick in Glasgow.

Gilbertfield is an interesting old mansion standing about two miles south-east of Cambuslang. It was built in 1607 on a simple L plan and was designed both for

Provand's Lordship. Oldest house in Glasgow

comfort and strength. For many years it was the residence of William Hamilton, of Gilbertfield, the friend and correspondent of Allan Ramsay.

Part of Bedlay House near Chryston dates from about the end of the sixteenth century. It is a quadrangular structure with high-peaked, crow-stepped gables, and two round turrets. It once belonged to the Earls of Kilmarnock, the last of whom was executed after the rebellion of the '45.

Hamilton Palace lies between the town and the river Clyde. The first castle of the Hamiltons in this vicinity was the royal castle of Cadzow, obtained from Robert the Bruce. Queen Mary resided there before the battle of Langside, and the Regent Murray therefore burnt the castle to the ground after the battle. Thirteen years later, in 1581, a new castle was erected on the site of the modern palace. The oldest part of the present building dates from this time, but large additions were made in 1705. Further buildings were added in 1822 by the tenth duke, and the structure is now a most imposing one. The interior is as grand as the exterior, and until recently contained the finest collection of art treasures in the country, but most of these were sold in 1882. The grounds are magnificent, and include Cadzow Forest with the interesting white cattle. The present owner is the thirteenth Duke of Hamilton.

Douglas Castle stands on the site of the "Castle Dangerous" of Scott, of which no trace remains. Adjoining the present building is a tower of an old castle, but considerably posterior in date to the historic pile burnt by the Good Lord James. After the conflagration of 1755, the last Duke of Douglas kept before him the prophecy that as often as Douglas Castle should be razed it should rise from the ruins with increased splendour. Accordingly in 1762 the present magnificent mansion was erected, and it is only a fraction of what the Duke intended it to be. The grounds and surroundings are particularly fine. The

Hamilton Palace

estate now belongs to the Earl of Home, a descendant of the nephew of the last duke, who died childless in 1761.

Mauldslie Castle is situated three miles from Carluke on the north bank of the Clyde. It is a large and imposing building with round, flanking turrets and a massive square tower. It was built for the fifth Earl of Hyndford in 1793, but its present owner is Lord Newlands. Mauldslie was originally a royal forest comprising most of the present parish of Carluke, but it was gifted in portions to several nobles by Robert the Bruce. The father of Lord Newlands acquired part of the estate some time after the death of the last Earl of Hyndford.

19. Communications — Past and Present.

Since the beginning of history Clydesdale has been one of the main routes between England and the Lowlands of Scotland. The barrier to communication is the Southern Uplands. But this area of high ground is deeply scored by the Clyde and the Annan, and the head waters of these streams come very near each other. They are joined by a pass a thousand feet above sea-level, and therefore from the earliest times to the present we find a tide of traffic ebbing and flowing along this channel. Two thousand years ago a Roman road was built along this route; to-day the London express passes over Beattock Summit and roars down Clydesdale within a few hundred feet of the old road by which the Roman legions marched.

Again, Scotland is more easily crossed from the Clyde to the Forth than at any other place, and therefore we find communication established between Glasgow and the east coast for many centuries. Glasgow stands at the convergence of many of the natural routes of the west of Scotland. The two already mentioned cross at Glasgow. The fertile Ayrshire plain is joined to the Clyde basin by a low valley that runs past Beith, Lochwinnoch and Paisley and points towards Glasgow. A second easy route from Ayrshire by the gap in the lava hills at Neilston and Barrhead is directed straight towards Glasgow. The western Highlands are practically shut off from the Lowlands by land, but they are easily accessible by sea, and the natural route from the west is up the Firth of Clyde and so into the heart of the Lowlands at Glasgow.

Clydesdale is connected with Tweeddale by the low valley of Biggar, and with eastern Ayrshire by a more difficult route up the Avon Water and another up the Douglas Water. Man has taken advantage of all these natural thoroughfares, and therefore we find both roads and railways running along these routes. A glance at the physical map on the cover will show at once how the valleys have controlled the directions of the roads and railways.

Although communication in Lanarkshire has been kept up for many centuries along the routes indicated, yet proper roads are of comparatively recent origin. In former times wheeled traffic was hardly possible, and most of the trade was done by pack-horses. It was not till the middle of the eighteenth century that a stage-coach ran between Edinburgh and Glasgow, and its average rate was less than four miles per hour. There was no coach communication between Glasgow and London till within 12 years of the nineteenth century. Until 1797 the letters to Glasgow were carried by a post-boy on horseback. Before the Clyde was deepened so that boats could come to Glasgow, the goods were carried from Dumbarton or Port Glasgow on pack-horses. The state of the Glasgow streets and roads may be imagined when we know that until 1777 only two men were employed in keeping up the "streets, causeways, vennels and lanes, the highways and roads, within and about the city, and territories thereof."

The passing of the Turnpike Roads Act in 1751 marked the beginning of a new era. In Lanarkshire Telford's Glasgow to Carlisle road became a model for future engineers. One of the most remarkable of Telford's achievements was the carrying of the road over Cartland Crags by a viaduct 130 feet high. The main artery of the county may be considered the road from Glasgow up the Clyde valley. Of the greatest importance also are the main roads from Glasgow to Edinburgh, one passing by Bathgate, the other by Shotts and Midcalder. The latter road is joined at Midcalder by the highway from Lanark to Edinburgh. Below Glasgow a highroad runs on each side of the Clyde connecting Lanarkshire with the most important towns of Dumbarton on the north and Renfrew on the south. The county is joined to Ayrshire by highroads running up the valley of the

Avon and the valley of the Douglas. Nithsdale may be reached from the Clyde by a road from Elvanfoot over Leadhills and Wanlockhead, or by a road up the Powtrail Water and over the Dalveen Pass. Eastwards Tweeddale can be reached easily by the route from Symington along Biggar Water to Peebles. Minor roads form such a complicated network as to baffle description.

The main railway lines form almost a duplication of the most important roads, and the reason for this has already been indicated. There are certain natural routes through the county, and the work of the engineer of both road and railway has consisted largely in taking advantage of these easy routes. The first railway in Lanarkshire to give communication from one end of the county to another was the Caledonian Railway opened in 1847 from Carlisle to Glasgow and Edinburgh, the divergence taking place at Carstairs. The line crosses Beattock Summit from Annandale and joins the Clyde at Elvanfoot. From this point for a long way, railway, road and river go side by side, crossing and recrossing like the strands of a cord. From Carstairs the main line passes through Wishaw, Motherwell, Uddingston and Cambuslang to Glasgow. The Caledonian main line to Edinburgh goes up the Clyde valley to Bellshill, then crosses the moors by Shotts and enters Linlithgowshire just west of Fauldhouse. The North British main line takes an easier route to the north by Falkirk, Polmont and Linlithgow, entering Edinburgh from the west. Another Edinburgh line passes by Coatbridge and Bathgate.

Main lines run west down the Clyde on both sides of the river. The two roads already mentioned from Lanarkshire to Ayrshire by Avondale and Douglasdale are accompanied by railway lines, and a line from Elvanfoot goes towards Nithsdale as far as Wanlockhead. The easy route from Clydesdale into the Tweed basin by the Biggar valley is utilised by the Caledonian line to Peebles. The dense population and the heavy mineral and goods traffic in the lower and middle wards have necessitated the building of almost innumerable lines of railways, lying over the lower part of the county like a spider's web of which the centre is at Glasgow.

A short portion of the Forth and Clyde canal lies in Lanarkshire. It passes through the north-west corner of the county taking advantage of the Kelvin valley. The Monkland Canal lies wholly in Lanarkshire. It starts from a branch of the Forth and Clyde canal at Port Dundas, Glasgow, runs east into the parish of old Monkland, passes through Coatbridge and terminates in the North Calder Water. The project was suggested in 1769, in order to ensure for all time a plentiful supply of coal to Glasgow. The canal was surveyed by James Watt and after some difficulties, chiefly of finance, was finished in the last decade of the eighteenth century. It was taken over by the Caledonian Railway along with the Forth and Clyde canal in 1867.

The palmy days of canal traffic both for passengers and goods have passed away. As railways were extended the importance of canals declined. The complete explanation of this is by no means easy. It has been attributed to their passing into the control of railway companies, but this explanation is not satisfactory. The smallness of the vessels in use and the consequent additional handling of goods undoubtedly militate against the greater use of canals in these days, when the whole tendency is to handle and carry goods in as large amounts as possible. With the adoption of improved methods of traction there seems no good reason why the importance of canal traffic should not to some extent be restored.

20. Administration and Divisions.

The county of Lanarkshire as an administrative unit probably dates from the time of David I, when the sheriffdom was inaugurated. At first the position of sheriff was hereditary and was held by one of the powerful families of the county. The Douglas family held the office for some time, for a period also it formed one of the hereditary titles of the Hamiltons. It was not till 1747 that appointments to the office were made in the modern method. Hamilton of Wishaw tells us that the sheriffdom originally included Renfrewshire until the two were separated by King Robert III in 1402. The county was originally divided into two wards, an upper ward, the capital of which was Lanark, and a lower ward, the capital of which was Rutherglen. In the middle of the eighteenth century the county was divided into an upper, a middle and a lower ward, the head towns of which were respectively Lanark, Hamilton and Glasgow. Since that time the middle ward has again been subdivided for administrative purposes, the chief towns being Hamilton and Airdrie.

The county possesses a lord-lieutenant, a vice-lieutenant, and a large number of deputy-lieutenants and justices of the peace, but the most important administrative body is the county council, which is composed of 70 elected members. Glasgow of course is not included in the county, as in 1893 it was constituted a county of a city with the lord-provost of the city as lord-lieutenant. The law is administered by a sheriff-principal and five sheriff-substitutes for general work along with two resident substitutes. The police force is a county constabulary, except in the large towns, most of which have their own separate forces.

The upper ward contains the following parishes: Biggar, Carluke, Carmichael, Carnwath, Carstairs, Covington, Crawford, Crawfordjohn, Culter, Dolphinton, Douglas, Dunsyre, Lamington and Wandel, Lanark, Lesmahagow, Liberton, Pettinain, Symington, Walston, and Wiston and Roberton. The middle ward contains the parishes of Avondale, Blantyre, Bothwell, Cambusnethan, Dalserf, Dalziel, East Kilbride, Glassford, Hamilton, New Monkland, Old Monkland, Shotts, and Stonehouse. In the lower ward are the parishes of Cadder, Cambuslang, Carmunnock, Cathcart (part), Eastwood (part), Govan, Rutherglen, and the parishes of Glasgow.

The county is represented by six members of Parliament, the districts being North-east Lanark, North-west Lanark, Mid-Lanark, South Lanark, Govan, and Partick.

In addition Glasgow returns seven members, and the burghs of Rutherglen, Hamilton, Airdrie, and Lanark unite with burghs outside the county in returning two more. Rutherglen has a share in the member for Kilmarnock Burghs, and the other towns mentioned combine with Falkirk and Linlithgow in returning the member for Falkirk Burghs.

The County Councils were established in 1889, and look after the finances, the roads and bridges, the public health, and the general administration. The parish councils administer the poor-laws, the unit of poor-law administration being the parish. The control of the insane is vested primarily in the Commissioners of Lunacy, and for each county there is a Lunacy Board. There are, finally, a number of burghs largely independent of the County Council. The burghs of Lanarkshire are Airdrie, Biggar, Coatbridge, Glasgow, Govan, Hamilton, Lanark, Motherwell, Partick, Rutherglen and Wishaw, and of these Glasgow, Lanark, and Rutherglen are royal burghs.

The burghs are managed by town councils, which administer the property of the burghs, impose the rates necessary for upkeep, and make bye-laws for the regulation of the trade of the town and the conduct of the inhabitants. Town councillors are elected for three years and one-third of the council retires annually. The councillors elect among themselves magistrates, who, besides performing other duties, act as judges in the cases that come before the ordinary police-courts.

It must not be forgotten that there is still a considerable amount of overlapping and confusion in the administrative divisions, not only of Lanarkshire but of all the counties of Scotland. The registration county is not the same as the civil county; the ecclesiastical parish differs from the civil parish; the district under municipal authority has no fixed relation to any of these areas. For example, a householder in the west of Glasgow may help to elect a town councillor for Glasgow, may vote for the School Board of Govan; and may have as his parliamentary representative the member for Partick, although the three towns are municipally absolutely distinct.

In 1889, under a Local Government Act, the Boundary Commissioners rectified some of the most glaring anomalies, and transferred certain areas from one parish to another, and in other cases from one county to another. The ecclesiastical divisions, however, in many cases still fail to harmonise with the civil divisions. For example, several of the parishes of Lanarkshire are in the synod of Lothian and Tweeddale, while the rest are in the synod of Glasgow and Ayr.

Since the Education Act of 1872, the management of education in Scotland has been entrusted mainly to School Boards, of which Lanarkshire has 49. Education is compulsory for children between the ages of five and 14 years, and is free to all. Above the primary schools there are two classes of higher schools, called intermediate and secondary. The former schools provide a three years' course, and the latter at least a five years' course of education after the elementary stages. Pupils who have passed through a secondary school with credit are quite able to go with profit directly to the University.

The County Council is also interested in secondary education, and is empowered to give grants to schools and to assist pupils by bursaries or otherwise. The two Glasgow training colleges for teachers, which were for so many years managed by committees of the Established Church and the Free (later United Free) Church, have now passed into the hands of a provincial committee, elected or nominated by various representative bodies.

21. The Roll of Honour.

In addition to one or two superlatively great men, Lanarkshire has produced a large number of persons who fall just short of that class, but whose names are nevertheless household words. There are at least two names that might legitimately be placed in the first of these classes. In the records of exploration no name stands higher than that of David Livingstone, and in the history of British physical science Newton alone takes rank before Lord Kelvin. It is a matter for further satisfaction to think that each of them was not only a genius in his own line, but also a genuine benefactor of mankind. Kelvin was not born in Lanarkshire, but his whole education and life-work were bound up inseparably with Glasgow University.

We can hardly claim for Lanarkshire any of the giants of pure literature, although the number of well-known names is very large. Nor, strange to say, have the many and exquisite beauties of Clydesdale inspired any literature of the first rank. One of the earliest of the county's famous writers was William Lithgow, who was born at Lanark in 1582. He travelled on foot all over Europe and North Africa, covering more than 36,000 miles, and on his return composed his best-known work, *The Rare Adventures and Paineful Peregrinations of William Lithgow*. He died in his native town about 1645. Allan Ramsay was born in 1686 at Leadhills, where his father was mine manager. He describes himself as

> "Of Crauford-muir, born in Leadhill
> Where mineral springs Glengonir fill
> Which joins sweet flowing Clyde."

Ramsay was 15 years of age when he left Leadhills, and his career afterwards was bound up with Edinburgh.

John Wilson was born at Lesmahagow in 1720. He was parish schoolmaster at Lesmahagow, Rutherglen, and Greenock. The last appointment he obtained only by promising to abandon "the profane and unprofitable art of poem-making." His best-known work is *The Clyde*, a long poem of nearly 2000 lines, of considerable merit and containing many interesting descriptions. Smollett was born in Dumbartonshire, but Glasgow has many claims on him. He was apprenticed to a Glasgow apothecary, and he drew on his Glasgow experiences for many of the characters in his novels. Adam Smith was not a native of Lanarkshire, but his connection with Glasgow University forms a strong link binding him to the county. He was educated at the college in Glasgow, and came back in 1751 to fill in succession the chairs of Logic and Moral Philosophy before retiring to his native

Kirkcaldy to produce his great work, *The Wealth of Nations*. On his election as Lord Rector of Glasgow University, he wrote "No man can owe greater obligations to a society than I do to the University of Glasgow."

The author of *Ye Mariners of England* was born in Glasgow in the vicinity of the High Street, and was educated at Glasgow University. All his life Campbell had a strong affection for his birth-place. "I was better pleased," he says, "to look on the kirk-steeples and whinstone causeways of Glasgow than on all the eagles and red deer in the Highlands." In the height of his fame he was three times elected Lord Rector of Glasgow University. Lanarkshire has a claim on De Quincey, from the fact that he resided in Glasgow between 1841 and 1843, and later on for shorter periods. He had lodgings in Rottenrow and Renfield Street. His greatest friends in Glasgow were Lushington, the professor of Greek, and Nichol, the professor of Astronomy, for De Quincey was at this time intensely interested in the latter subject.

John Gibson Lockhart, the biographer of Scott, was a Lanarkshire man, born in Cambusnethan manse, near Wishaw. His life-work, however, was done in Edinburgh and London. He occasionally collaborated with Christopher North in the *Noctes Ambrosianae*; and the latter, too, had close associations with Glasgow, as he was educated at the University, and always looked back to the time he spent there as the happiest of his life. Sheridan Knowles, the dramatist, was for some years a teacher of elocution in the Trongate, Glasgow. His best-known play *Virginius* was produced for the first time in the Theatre Royal, Glasgow, in 1820, and its success there induced Macready to stage it at Covent Garden, London, where it immediately made the author famous.

Thomas Campbell

Sir Archibald Alison, author of *The History of Europe*, was appointed Sheriff of Lanarkshire in 1834, and thereafter resided at Possil House, Glasgow. He was elected Lord Rector of the University in succession to Macaulay. Michael Scott, the author of *Tom Cringle's Log* and *The*

Cruise of the Midge, was a Glasgow man, having been born at Cowlairs. He was buried in Glasgow Necropolis. The kindly Dr John Brown, author of *Rab and his Friends*, was another Lanarkshire man who gravitated to Edinburgh. He was born in his father's manse at Biggar, but when he was eleven years of age the family removed to the capital, which was his home for the rest of his life.

Alexander Smith, the author of *Dreamthorp* and *A Summer in Skye*, was born at Kilmarnock, but was educated and went to work in Glasgow, and it was in this city that he began to contribute his poems to the local press. Smith knew Glasgow thoroughly, and one of his verses on the city is worth quoting.

> "Draw thy fierce streams of blinding ore,
> Smite on a thousand anvils, roar
> Down to the harbour-bars;
> Smoulder in smoky sunsets, flare
> On rainy nights, with street and square
> Lie empty to the stars.
> From terrace proud to alley base
> I know thee as my mother's face."

William Black, the popular novelist, was born in Glasgow in a house in the Trongate, and before going to London he wrote for a time for *The Glasgow Weekly Citizen*. Robert Buchanan and he were close friends in Glasgow, where both determined some day to be famous. Though the link is much slighter, it is worth recalling the fact that James Boswell was educated partly at Glasgow University. It was at the famous Saracen's

Head inn that he and Dr Johnson stayed when passing through Glasgow on their tour in Scotland.

In science the two most illustrious names connected with Lanarkshire are James Watt and William Thomson, Lord Kelvin. Neither was born in Lanarkshire, but the life work of each was done in Glasgow. James Watt came from Greenock in 1754, and found employment in the little shop of a mechanic calling himself an "optician." Here Professor Anderson handed him the model of the Newcomen engine to mend, and so originated one of the greatest discoveries in the history of the world. It was on Glasgow Green that Watt was walking one Sunday afternoon in 1765, pondering over his engine, when just as he got to the herd's house the "idea of a steam condenser flashed upon his mind."

For 67 years Lord Kelvin lived in Glasgow, and by his long series of brilliant researches in every department of physical science caused the name of his university to be renowned throughout the civilised world.

William Cullen, the celebrated physician, was born in Hamilton and educated at Glasgow University. He was the first holder of the chair of medicine there. The medical faculties of Scottish universities have always been particularly strong. Two of their most famous sons, William and John Hunter, were born in Lanarkshire at Long Calderwood, near East Kilbride. William Hunter was the first professor of anatomy to the Royal Academy, London. His magnificent collections in literature and science were bequeathed to Glasgow University, where they form the most important part of the Hunterian

Museum. Another distinguished holder of a medical chair was Joseph Black, the famous chemist. He was born in France, but for ten years he was professor of medicine in Glasgow Univerity. He showed the distinction between air and carbon dioxide ("fixed air" it was then called) and made the great discovery of latent heat.

It is now generally admitted that William Symington was the first man successfully to originate steam navigation. He was born in the village of Leadhills, and his little whitewashed cottage can still be seen there, with the fine obelisk behind it erected to his memory. Of him Lord Kelvin said, "Symington was the real discoverer and the practical originator, the engineer who foresaw that good was to be done, who understood how to make the machine to do it, and who had the true engineering and mechanical principle for doing it in the right way." In 1802, several years before Bell or Fulton was successful, Symington's stern-wheel steamer the "Charlotte Dundas" gave a successful demonstration on the Forth and Clyde canal.

Sir A. C. Ramsay, one of the most brilliant of the many famous geologists that Scotland has produced, was a Glasgow man; and it was at the British Association meeting in Glasgow, in 1840, that he first attracted the attention of men of science to his work. He afterwards became Director-General of the Geological Survey of Great Britain.

Lanarkshire has taken such a prominent place in the industrial history of the country, that we might expect that in the field of applied science her sons would be in the van, and this we find to be the case. In the mining and metal industries particularly, there are several names of more than local reputation. The smelting of iron was revolutionised by the introduction of the hot blast, suggested by James B. Neilson, manager of the Glasgow gasworks. It was David Mushet again who in 1801 pointed out that the miners were throwing away under the name of "wild coal" a very valuable iron ore, the blackband ironstone. Mushet was born near Edinburgh, but was employed in the Clyde iron works. By dint of severe application he became one of the foremost authorities on iron and steel in the country.

James Young, the founder of the important oil-shale industry of Scotland, was born in Glasgow in 1811. He was an assistant in the old Andersonian College, Glasgow, and it was there he obtained the knowledge of minerals that enabled him to create the paraffin industry of this country. Another famous Glasgow chemist was Charles Macintosh. While experimenting with the naphtha obtained by the distillation of coal-tar, he discovered a method of dissolving india-rubber and thus of making cloth waterproof. The invention was patented in 1823. MacArthur, one of the discoverers of the method of recovering finely divided gold by the cyanide process, was a Glasgow man.

The name of David Livingstone will always stand for what is best in the Scottish character. He combined determination to succeed, eagerness for learning, and nobility of mind with natural aptitude for his life-work to such a degree, that we may say without exaggeration that his gifts as an explorer rose to the height of genius. The house in Blantyre where he was born still stands, although the mill where he worked and snatched a

David Livingstone

sentence from an open book as he passed to and fro is now dismantled. Like Young, he obtained his early scientific education in the Andersonian College.

The two most famous soldiers associated with the county are Sir John Moore and Colin Campbell, Lord

Clyde. The former was born in Glasgow just east of the Candleriggs, the latter in High John Street off George Street. Another distinguished Lanarkshire soldier was General Roy, born in Carluke parish. His fame, how-

Sir Colin Campbell

ever, rests chiefly on his work as a surveyor. He was one of the early pioneers of that magnificent organisation, the Ordnance Survey, and wrote also a standard work on Roman Military Antiquities in Britain. The famous General Wolfe lived for some time in Glasgow at Camlachie House. He was not favourably impressed by the citizens, for he described them as "excessive blockheads, so truly and obstinately dull that they shut out knowledge at every entrance."

The names of some well-known divines are linked with Glasgow. Bishop Elphinstone, the founder of King's College, Aberdeen, was a Glasgow man. He became Bishop of Aberdeen and was largely instrumental in introducing printing into Scotland. It is supposed that John Knox studied at the university, although there is considerable doubt on this point. The famous Andrew Melville was Principal immediately after the Reformation. Zachary Boyd was minister of the cathedral when Cromwell visited Glasgow, and so soundly rated the General that he narrowly escaped being pistolled. Dr Chalmers lived in Charlotte Street near Glasgow Green, and his famous astronomical discourses were delivered in the Tron Kirk. The house of his hardly less famous assistant, Edward Irving, was in Kent Street, a stone's throw from his own dwelling.

22. THE CHIEF TOWNS AND VILLAGES OF LANARKSHIRE.

The figures in brackets after each name give the population in 1901, and those at the end of each section are references to the pages in the text.

Airdrie (22,288), Gaelic **ard ruith**, high pasture run, was a mere hamlet until the rise of the coal and iron industry made it a large and prosperous town. Some historians believe it to be the site of the great battle between the Pagans and the Christians of Strathclyde, in which the pagan chief was slain and his bard Merlin forced to flee, while the victor recalled the good St Mungo back to Glasgow from his exile in Wales. It stands on the highroad between Edinburgh and Glasgow, and is connected to the latter town by the Monkland Canal. There are several important coal and iron mines in the vicinity. In addition to foundries and engineering shops there are works for calico-printing, paper-making, cotton and wool manufacturing, oil-refining and fire-clay making. Airdrie was the first town in Scotland to adopt the Free Library Act of 1856. Its academy has a high reputation. (pp. 26, 33, 55, 137, 138.)

Baillieston (3784) perhaps takes its name from the "bailie" who managed the estate of Monkland for the monks of Newbattle, or from the Baillie family, prebendaries of Glasgow Cathedral. It is a little coal mining town just over six miles east of Glasgow. A confectionery work and large nurseries give employment to numbers of the inhabitants. (p. 50.)

Bellshill (8786) is a mining town in Bothwell parish, nine miles from Glasgow. The coal and iron industries are its chief support. It is a long straggling town, with a first-class school, Bellshill Academy. (p. 134.)

Biggar (1366), Norse—**bygg garðr,** barley-field, is a little town on Biggar Water, a tributary of the Tweed. It is an old historic town and has been a burgh since the fifteenth century. Its parish church was founded in 1545. In its churchyard lie the

"The Cadger's Bridge," Biggar

Gledstanes of Libberton, the ancestors of William Ewart Gladstone. In the vicinity was Boghill Castle (now destroyed) the seat of the powerful Flemings, who settled here in the twelfth century. Blind Harry tells of a great battle fought on Biggar Moss in which Wallace and his men slew 11,000 Englishmen. The Cadger's Bridge is supposed to have obtained its name from the tradition that Sir William Wallace crossed it in the disguise of a cadger in order to reconnoitre the English camp. (pp. 12, 14, 17, 31, 55, 58, 106, 111, 120, 132, 138, 144.)

Blantyre (2521) stands on the right bank of the Rotten Calder, just over six miles from Glasgow. It is a mining town, and is surrounded by coal and iron workings. The town is famous for having been the birthplace of David Livingstone. In the vicinity is Blantyre Priory. (pp. 24, 56, 123, 148.)

Bothwell (5179) stands on the right bank of the Clyde, about six miles to the south-east of Glasgow. The charming surroundings have made it a popular residential town. There are

"Roman" Bridge near Bothwell

ruins of an old church founded in 1398 by Archibald the Grim, Earl of Douglas. Joanna Baillie, the poetess and correspondent of Scott, was born at the manse in 1762. Bothwell Brig was the scene of the defeat of the Covenanters in 1679. Across the Calder there is an old single-span bridge, believed to be of Roman workmanship. The lands of Bothwell were originally possessed by the Murrays, from whom they went to the Douglas family in 1361. Bothwell passed out of their possession for a time, but was regained by Archibald Douglas, fifth Earl of Angus. The present proprietor is the Earl of Home, whose mother was heiress of the last Lord Douglas. (pp. 11, 22, 24, 39, 55, 65, 102, 119, 121.)

Calderbank (2077) is an industrial village two miles from Airdrie. It depends to a large extent on the finely equipped steel-works of the Calderbank Company.

Cambuslang (12,252) is prettily situated on rising ground about three miles from Glasgow. There are several coal-pits in the neighbourhood and the Steel Company of Scotland has large works at Newton. Of late years it has become popular as a residential suburb of Glasgow. Above the town is Dechmont Hill, where in pagan times the Beltane fires were lit. The town is known for the famous "Cambuslang Wark," which took place in 1742—a gigantic religious revival. (pp. 56, 128, 134.)

Carfin (2115) is a large village a little less than two miles north-east of Motherwell, of which it may be considered almost a suburb. The inhabitants are chiefly colliers engaged in the neighbouring pits.

Carluke (4740) stands on the right bank of the Clyde just over two miles from the river, and about twenty miles from Glasgow. Its situation is magnificent, overlooking all the lower basin of the Clyde. The fruit-growing industry is carried on all round it, and there are jam factories in the town. In the seventeenth century it was quite an important place, but later it declined greatly. In the last century it has prospered largely owing to the neighbourhood of mines. Several valuable minerals are worked in the vicinity. At the present time its advantages as a residential place are recommending it to many Glasgow business men. (pp. 33, 35, 40, 44, 55, 131.)

Chapelhall (2030) adjoins Calderbank and, like it, is dependent on the steel works and coal-pits of the Calderbank Company.

Cleland (with **Omoa**) (2729) is a large village three miles to the east of Motherwell. It is engaged chiefly in coal-mining.

Coatbridge (36,991), just two miles from Airdrie, has sprung into importance owing to the development of the coal and iron in the vicinity. It is the iron-smelting town of Lanarkshire, and has in its neighbourhood more blast-furnaces than any other town in Scotland. In addition to the making of pig-iron there are malleable iron works, steel works and rolling mills. Other industries are the making of boilers, tubes, heavy metal goods of every description, ordinary bricks and fire-bricks. To the west of the town lived the blind poetess, Janet Hamilton. The Coatbridge Technical School is an important centre of industrial education. (pp. 26, 33, 71, 134, 135, 138.)

Glasgow (761,709) is by far the largest and most important commercial town in Scotland, and in point of size is the second city of the kingdom. Its rapid growth has been due largely to its favourable position. Before the coming of the Romans much of the present site of Glasgow must have been under water, for canoes have been found buried in the silt hundreds of feet away from the present banks of the river. The position of the end of the Roman wall, however, shows that by that time the level of the land stood approximately the same as now.

The real beginning of Glasgow may be correlated with St Kentigern, generally called St Mungo (loved-one or blessed). He settled by the banks of the Molendinar, and here took place his meeting with St Columba about 584. Then for five hundred years the record of the little place is a blank, till the see was restored in 1115 with John, the first of the new bishops, who replaced the early church by a new cathedral. By a charter of William the Lion, Glasgow was constituted a burgh with the privilege of holding a weekly market and an annual eight days' fair. On the burning of the cathedral the present crypt was built in 1197. The first permanent bridge over the Clyde was built by Bishop Rae about 1350. It had eight arches and was solidly constructed of stone.

The immediate effect of the Reformation on the material welfare of Glasgow was almost disastrous. The loss to the town of the wealthy clergy and university students was severe. Yet by turning the energies of the citizens away from ecclesiastical matters, in which hitherto they had exclusively been exercised, and by forcing the city to look round for other openings for industry,

Jamaica Bridge

the effect of the Reformation in the long run was extremely beneficial.

Glasgow entered into the Darien scheme with great enthusiasm; the council invested £3000, many of the citizens subscribed largely and not a few accompanied the expedition. The failure of the scheme brought ruin to many families in Glasgow, and was one cause of the bitter enmity of the city to the Union of the Kingdoms. It seems strange to read that at this time the city was noted for its beauty, being reckoned superior even to Edinburgh. Defoe states, "Glasgow is the beautifullest little City I have seen in Britain." Another description was "a much sweeter and more delyghtful place than Edinburgh"; and a third, "the nonsuch of Scotland."

It was not till the eighteenth century, however, that Glasgow was of any importance as a commercial centre. At the beginning

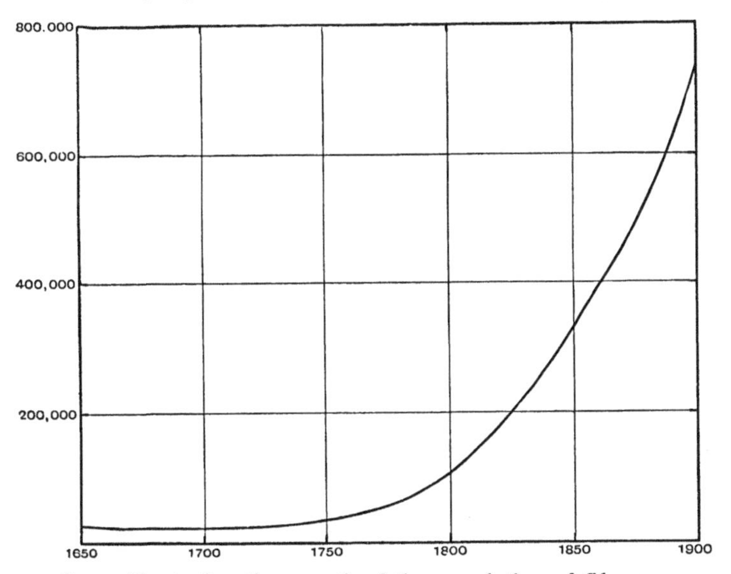

Curve illustrating the growth of the population of Glasgow from 1650 to 1900

of the century its population was little more than 12,500, and the total tonnage of its vessels barely more than 1000 tons. It was the exploiting of the coal and iron wealth of Lanarkshire, however, that brought Glasgow quite into the front rank of British cities. The description of the growth of the manufactures of the county in a former chapter may be taken to apply to Glasgow. The growth of the city is well illustrated by the curve of population shown on p. 157. The rise in the eighteenth century after the Union in 1707 is clearly seen, and so also is the sudden increase in population towards the end of the eighteenth and the beginning of the nineteenth centuries, due to the industrial revolution.

Royal Exchange, Glasgow

Glasgow is undoubtedly a well-built city. The material used is chiefly yellow or white sandstone obtained largely from local quarries. Of late years the red freestone of Dumfriesshire has become extremely fashionable. In addition to the Cathedral, the University and the Municipal Buildings, described elsewhere, mention must be made of the new Art Galleries and Museum in Kelvingrove. The building has not only a handsome exterior but possesses one of the finest collections of paintings in this country.

Glasgow has a high reputation in municipal matters. Its water supply from Loch Katrine is not only ample but of exceptional purity, its electric car service is a model to the world, and its new sewage schemes are on a gigantic scale. Yet strange to say it has made no serious attempt to purify its atmosphere. Undoubtedly one of the most attractive features of Glasgow is the ease with which one can get out of it. In less than two hours one can be among some of the grandest and wildest scenery of the West Highlands, and this can be effected with a comfort, cheapness and celerity probably not equalled by any other town of its rank in Britain.

In 1450 Bishop Turnbull obtained from Pope Nicholas V a bull authorising the erection of a university in Glasgow. At first the classes were held in the Cathedral, but afterwards in a building in Rottenrow. In 1459 the University received ground on the east side of High Street, on which, in the seventeenth century, were erected the old University buildings.

Like the Cathedral, the very existence of the University was threatened by the Reformation, for its officers were of course all churchmen. In 1571 the number of students on the roll was about a dozen and the annual revenue about £25. Andrew Melville became Principal in 1574, and by his exertions the institution was put on a sounder footing.

In 1677 the University received one of its most important foundations, the Snell Exhibitions, the winners of which still hold their scholarships at Balliol College, Oxford. Adam Smith and Sir William Hamilton were Snell Exhibitioners. By the middle of the nineteenth century the old buildings had become quite unsuitable. In 1864 the old buildings were sold and land was purchased at Gilmorehill, a magnificent site in the west end of the city. The new buildings were designed by Sir George Gilbert Scott, in the Early English style with Scottish modifications.

The Bute Hall was added by the Marquis of Bute, and the Randolph Hall by Charles Randolph, the shipbuilder. The

Glasgow University and Kelvingrove Park

Students' Union was erected by means of a bequest from Dr McIntyre in 1885. The University buildings are grouped round two quadrangles with a magnificent frontage to the south, culminating in a spire rising to a height of nearly 280 feet. Additional buildings have since been added, including two splendid separate departments for Botany and for Natural Philosophy, and the fine "James Watt" Engineering Laboratories.

The Library contains about 200,000 volumes. Several of the University's greatest literary treasures are housed in the Hunterian Museum, which possesses about 12,000 printed books and six or seven hundred manuscripts. There are nearly 500 examples of fifteenth century printing, including thirteen volumes printed by Caxton. Other treasures are an illuminated manuscript Psalter of the twelfth century, and two beautiful manuscripts of Chaucer's *Romaunt of the Rose*, one of which is the finest extant. Then there are first editions of Milton's *Paradise Lost* and Spenser's *Faerie Queene*; a First Folio Shakespeare; and some beautiful examples of binding from the libraries of Diana of Poitiers, Louis XIII of France, and other royal book-lovers. The collection of coins and medals is unrivalled.

Affiliated to the University is Queen Margaret College for women students, instituted by the munificence of Mrs John Elder of Govan. The Faculties of Medicine, Arts and Science in the University itself are open to women. The Glasgow and West of Scotland Technical College now rivals the University in numbers and equipment. In the huge new building almost every branch of science is pursued. The Chemical department has long had an especially high reputation; and particular attention is given to trades classes, such as printing, weaving, bootmaking and bakery.

Since 1885 Glasgow has returned seven members to parliament, the electoral divisions being Bridgeton, Camlachie, St Rollox, Central, College, Tradeston, and Blackfriars. (pp. 16, 23, 24, 25, 32, 33, 35, 36, 39, 42, 46, 48, 49, 50, 51, 53, 54, 55, 58, 61, 62, 68, 70, 71, 72, 73, 75, 77, 78, 79, 80, 84, 90, 91,

92, 93, 96, 98, 100, 101, 104, 108, 109, 111, 113, 115, 116, 118, 120, 126, 127, 132, 133, 134, 135, 136, 137, 138, 139, 140, 141, 142, 143, 144, 145, 146, 147, 149, 150.)

Govan (82,174) is really now a part of Glasgow, although it still remains a separate municipality. It stretches along the south side of the Clyde to the west of Glasgow. At the end of the eighteenth century the population was probably less than 2000; and Govan remained a mere village to the middle of the nineteenth century. It is now a great engineering and ship-building centre. From the famous Fairfield yard, founded by John Elder, whose father had been associated with Napier in making marine engines, more men-of-war have been launched than from any other yard on the Clyde. A fine park and library keep green the name of Elder. (pp. 6, 73, 77, 137, 138, 139.)

Hamilton (32,775) is prettily situated on the south bank of the Clyde among very picturesque surroundings and about ten miles south-east of Glasgow. There are few manufactories of any importance, although the town is in the heart of the coal-measures and many of the inhabitants are engaged in coal-mining. William Cullen the famous physician was born here in 1710. Hamilton Academy is one of the finest schools in Scotland. Near Hamilton are Cadzow Castle and Hamilton Palace. (pp. 55, 61, 98, 100, 101, 129, 136, 137, 138, 145.)

Holytown (4483), eleven miles from Glasgow, is in Bothwell parish, in the midst of a mineral working district.

Lanark (6567), a royal burgh since the time of David I, was the ancient capital of the county. It stands in a fine situation among beautiful scenery, and is a favourite residential town. William Lithgow was a native of Lanark and is buried there. New Lanark, a mile away, was the scene of Robert Owen's social experiments. Lanark is the best centre from which to see the Falls of Clyde, Cartland Crags and some of the best parts of the river valley. At the annual horse-races the Lanark Silver Bell is competed for. This is perhaps the oldest sporting trophy in existence: tradition describes it as a gift from William the Lion, although it probably does not date back quite so far. (pp. 19, 20, 21, 26, 55, 58, 65, 84, 98, 100, 111, 113, 114, 133, 138, 141.)

Larkhall (11,879) stands on the right bank of the Avon, not far from its junction with the Clyde. It was a small village until about 100 years ago. The inhabitants are engaged largely in mining and also in the bleaching industry.

High Street, and Wallace's Monument, Lanark

Mossend (3415) lies between Bellshill and Holytown, and owes its origin to the coal and iron industry. It contains the large iron and steel works of the Summerlee and Mossend Company.

Motherwell (30,418) takes its name from a well dedicated in early times to the Virgin Mary. The town is entirely a product of the nineteenth century, as on its site there was previously not even a village. It stands on the north bank of the Clyde just over a mile from the river and faces Hamilton. In the very centre of the coal-mining district, Motherwell is one of the most important towns in the kingdom for the manufacture of iron and steel. The Glasgow Iron and Steel Company, the Dalzell Steel Works, the Etna Iron and Steel Company, the Lanarkshire Steel Company (the two last mentioned at Flemington), and many others are engaged in the making of steel ingots, sheets, bars, rails, girders, boiler-plates and every conceivable article of steel. Roof and bridge work is an important branch, and another is the making of cranes. In the vicinity are also fire-clay works, boiler works, rivet and bolt works, and engineering works of every description. The town lies on the main Caledonian line to England and is an important railway centre. (pp. 55, 58, 60, 72, 73, 75, 88, 134, 138.)

Newarthill (2156) is a large village three miles to the east of Motherwell. Its inhabitants are engaged chiefly in the surrounding collieries.

Newmains (2755) is a little town in Cambusnethan parish, two miles north-east of Wishaw, engaged in coal-mining and iron-working. The Coltness Iron Company possess the most important works in the vicinity.

Newton (2139) is a little town just over five miles to the south-east of Glasgow. The Steel Company of Scotland have large works here; and there is also a large nail-factory. (p. 72.)

Partick (54,298) is now really a part of Glasgow, although it still remains a separate municipality. The town is very ancient and has had ecclesiastical associations with Glasgow for many centuries. In the twelfth century lands at Perdeyc were granted by David I to the Bishop of Glasgow; and later there was a bishop's residence at Perthik. Until sixty years ago Partick was a little country village, but the wave of industrialism extending from Glasgow has now completely penetrated and surrounded it. Its ship-building yards fringe the north bank of the Clyde, and the town continually resounds with the clatter of the riveters' hammers. There are extensive flour-mills near the Kelvin. At the west end of the burgh is Victoria Park, a public pleasure-ground that contains one unique feature. This is the Fossil Grove, a number of trunks and roots of fossil trees belonging to the Carboniferous System. (pp. 137, 138, 139.)

Strathaven Castle

Rutherglen (17,220) was for long the most important town in lower Lanarkshire. Even in 1402, when the county was divided into two wards, Rutherglen was considered the chief town of the lower ward. Its jurisdiction extended over Glasgow, and the Bishops of Glasgow had considerable difficulty in obtaining freedom from the exactions of the neighbouring town. It is said that Rutherglen was constituted a royal burgh in 1126 by David I; certainly a charter granted by Robert the Bruce quotes a confirmation of this by William the Lion. Part of the old church still remains, one of the few fragments of Norman architecture left in Lanarkshire. John Wilson, the author of the descriptive poem, *The Clyde*, was a teacher in Rutherglen for some time. Rutherglen stands at the western extremity of the coal-measures, and has large chemical works, tube works, rope works, dye works and brick works. (pp. 84, 94, 96, 101, 113, 126, 136, 138, 141.)

Shettleston (12,154), once a weaving village, is a little colliery town, three miles to the east of Glasgow. In recent years it has grown in favour as a residential suburb.

Stonehouse (2961), on the right bank of the Avon Water, stands high in a fine, healthy situation. It is a town of recent growth and was formerly a weaving village. It is situated near good seams of coal and limestone. (p. 33.)

Strathaven (4076) is a weaving town on the left bank of Avon. Placed as it is six hundred feet above sea-level and on the edge of the Lanarkshire moors, the locality is an ideal residential one. It is an old town that grew up under the protection of the Lords of Avondale, the ruins of whose castle are still conspicuous. (pp. 32, 124.)

Uddingston (7463) is prettily situated about seven miles east of Glasgow, of which it is largely a residential suburb. There is also an industrial population, chiefly miners. The town possesses jam and confectionery works. (pp. 109, 134.)

Wishaw (20,873), fifteen miles from Glasgow, although a large town to-day, was sixty years ago a mere village of 3000 inhabitants. It is one of the most important coal towns in the kingdom. Its industries also include iron and steel making, foundry work, railway-waggon building, and fire-clay making. (pp. 2, 134, 136, 138, 142.)

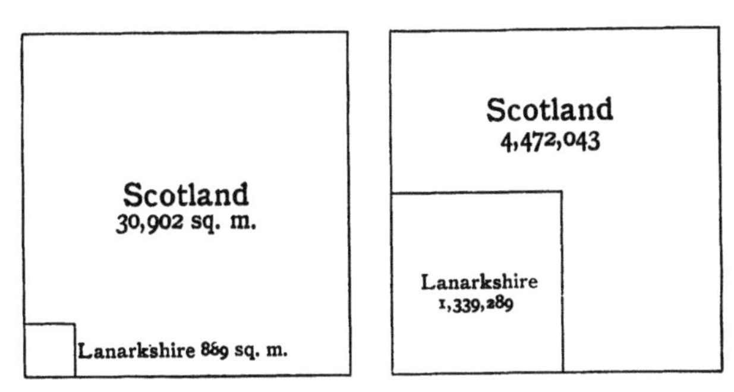

Fig. 1. Comparative areas of Lanarkshire and all Scotland

Fig. 2. Comparison in population of Lanarkshire and all Scotland

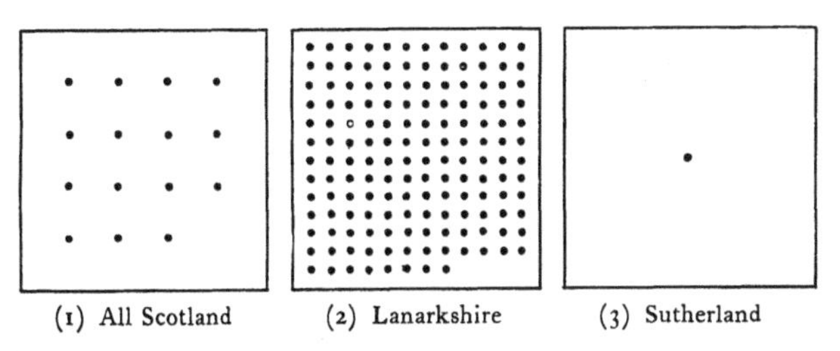

(*The dots represent the number of persons in each tenth of a square mile*)

Fig. 3. Density of population of all Scotland, Lanarkshire, and Sutherland

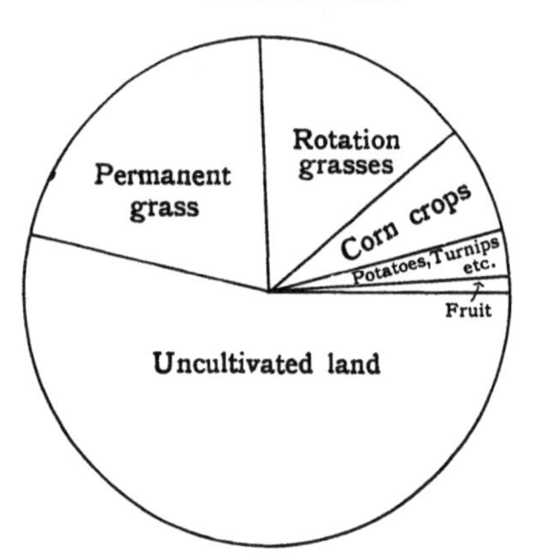

Fig. 4. Comparative areas under different crops in Lanarkshire

(*From Agricultural Returns*, 1908)

Cattle 71,636
Horses 8,800
Pigs 7,296
Sheep 257,779

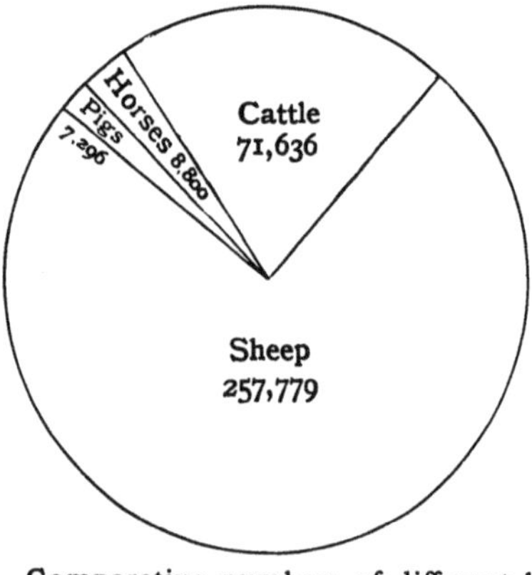

Fig. 5. Comparative numbers of different kinds of live stock in Lanarkshire

(*From Agricultural Returns*, 1908)

www.ingramcontent.com/pod-product-compliance
Ingram Content Group UK Ltd.
Pitfield, Milton Keynes, MK11 3LW, UK
UKHW042144280225
455719UK00001B/85